ISBN 978-1-331-91533-1
PIBN 10253421

This book is a reproduction of an important historical work. Forgotten Books uses state-of-the-art technology to digitally reconstruct the work, preserving the original format whilst repairing imperfections present in the aged copy. In rare cases, an imperfection in the original, such as a blemish or missing page, may be replicated in our edition. We do, however, repair the vast majority of imperfections successfully; any imperfections that remain are intentionally left to preserve the state of such historical works.

1 MONTH OF
FREE
READING

at

www.ForgottenBooks.com

By purchasing this book you are
eligible for one month membership to
ForgottenBooks.com, giving you
unlimited access to our entire
collection of over 700,000 titles via
our web site and mobile apps.

To claim your free month visit:

www.forgottenbooks.com/free253421

English
Français
Deutsche
Italiano
Español
Português

www.forgottenbooks.com

Mythology Photography **Fiction**
Fishing Christianity **Art** Cooking
Essays Buddhism Freemasonry
Medicine **Biology** Music **Ancient
Egypt** Evolution Carpentry Physics
Dance Geology **Mathematics** Fitness
Shakespeare **Folklore** Yoga Marketing
Confidence Immortality Biographies
Poetry **Psychology** Witchcraft
Electronics Chemistry History **Law**
Accounting **Philosophy** Anthropology
Alchemy Drama Quantum Mechanics
Atheism Sexual Health **Ancient History**
Entrepreneurship Languages Sport
Paleontology Needlework Islam
Metaphysics Investment Archaeology
Parenting Statistics Criminology
Motivational

THE PRIZE ESSAY

ON THE APPLICATION OF

RECENT INVENTIONS

COLLECTED AT THE

GREAT EXHIBITION OF 1851,

TO THE PURPOSES OF

PRACTICAL BANKING.

BY GRANVILLE SHARP.

(of Norwich)

"The Earth is the Lord's, and all that therein is ; the compass of the World and they that dwell therein."—*Psalms.*

LONDON:

REPRINTED FROM THE *BANKERS' MAGAZINE* OF JANUARY AND
FEBRUARY, 1852, BY WATERLOW & SONS, LONDON WALL.

1852.

PREFACE.

THE circumstances under which this Essay was written will be sufficiently explained by the following Extracts from the *Bankers' Magazine:*—

INDUSTRIAL EXHIBITION.—BANKING PRIZE ESSAY.—In our January number, we made the following announcement :—

" We are authorized to announce that J. W. Gilbart, Esq., F.R.S., will present the sum of ONE HUNDRED POUNDS to the author of the best essay which shall be written in reply to the following question :—

"*In what u ay can any of the articles collected at the Industrial Exhibition of 1851 be rendered especially serviceable to the interests of 'Practical Banking?'*

" These articles may be architectural models that may suggest improvements in the bank house or office—inventions by which light, heat, and ventilation may be secured, so as to promote the health and comfort of the bank clerks—discoveries in the fine arts, by which the interior of a bank may be decorated, or the bank furniture rendered more commodious—improvements in writing paper, pens, ink, account books, scales, letter copying machines, or other instruments used in carrying on the business—improvements in printing and engraving, by which bankers may get their notes, receipts, letters of credit and other documents of a better kind, at a less expense, or so as to prevent forgery—new inventions in the construction of locks, cashboxes, and safes, which shall render property more secure against fire or thieves—and generally all articles of every kind which can be so applied as to improve, cheapen, or facilitate any of the practical operations of banking."

We have since informed our readers that the following gentlemen have consented to act as adjudicators, viz :—

P. F. AIKEN, Esq., Managing Director of Stuckey's Banking Company, Bristol.

C. BROWN, Esq., a Director, and (the late) Manager of the Cumberland Union Bank, Workington.

P. M. JAMES, Esq., a Director, and the Manager of the Manchester and Salford Bank, Manchester.

We are now authorized by the adjudicators to publish the following regulations, for the information of those parties who are disposed to become candidates.

1st. The essays to be sent, post paid, to either of the adjudicators, by the 1st day of next November. Each essay to be marked " Banking Essay," and to have a motto or initial affixed, which must also be written on the outside of a sealed letter, containing the name and address of the author. The unsuccessful essays to be returned, when claimed, to their respective authors, with the accompanying letters unopened.

2nd. The copyright of the essay shall remain with the author, but he shall consent to the publication of the essay in the *Bankers' Magazine;* and Mr. Gilbart shall have the power, within six months after the adjudication, of printing any number he may think proper, at his own expense, for gratuitous distribution.

3rd. The adjudication will probably take place in the month of November, or at least in time for the successful essay to be published in the *Bankers'*

Magazine for January, 1852. The adjudicators shall not be expected to give any reasons for their preference, and their decision shall be final.—*Bankers' Magazine, May,* 1851.

THE BANKING PRIZE ESSAY.—We have the pleasure of presenting our readers this month with the first portion of the Banking Essay, to which has been awarded the prize of £100, offered by Mr. Gilbart.

Our readers are aware, from the notice in our May number, that the gentlemen who had consented to act as adjudicators were *Mr. Aiken, Mr. Brown,* and *Mr. James.* Unfortunately, when the time for receiving the essays had arrived, Mr. Brown had an attack of paralysis; Mr. Aiken and Mr. James therefore proceeded in the examination of the essays, upon the understanding that, should they differ in opinion with reference to the merits of the two best which might be sent in, these should be referred to an umpire. After carefully reading all the essays submitted to them, the adjudicators did differ in opinion as to which of two was entitled to the prize, and these two Essays were accordingly placed in the hands of *George Grote, Esq.,* who consented to act as umpire, and his adjudication is stated in the following letter addressed by him to Mr. Gilbart:—

12, Saville-row, December 24th, 1851.

MY DEAR SIR,—Pursuant to your request, I have perused and compared the two Essays sent to me on the question—"In what way can any of the articles collected at the Industrial Exhibition of 1851, be rendered especially serviceable to the interests of practical banking?" with reference to the prize offered by you.

As you have invited me to adjudicate on the comparative merits of the two, I beg to state that, in my opinion, the prize ought to be awarded to that Essay which bears the motto—.

"Every new development of Science may be looked upon as the opening of a fresh mine of wealth, awakening moral industry and sagacity, and employing, as it were, new capital of mind."

Respecting the other Essay, though I cannot award to it the prize, I feel bound to add that I consider it an able and creditable composition.

I remain, my dear Sir, yours very truly,

J. W. Gilbart, Esq.   (Signed)   GEORGE GROTE.

The writer of the Essay, to which the prize has been awarded by Mr. Grote, is Mr. Granville Sharp, an accountant in the East of England Bank, at Norwich.—*Bankers' Magazine, January,* 1852.

𝔓𝔯𝔦𝔷𝔢 𝔈𝔰𝔰𝔞𝔶

ON

THE ADAPTATION OF RECENT INVENTIONS

COLLECTED AT THE GREAT EXHIBITION OF 1851,

TO THE PURPOSES OF

PRACTICAL BANKING.

By GRANVILLE SHARP.

"Every new development of Science may be looked upon as the opening of a fresh mine of wealth, awakening moral industry and sagacity, and employing, as it were, new capital of mind."

PERHAPS there are few men (if there be any), since the days of Adam Smith, who have done so much to advance Banking, as a science, and to promote its cultivation and practical development, as James William Gilbart, F.R.S., the originator of this humble Essay, who has for many years presided over the Executive of the London and Westminster Banking Company, a commercial establishment, of recent origin indeed, but in the magnitude and importance of its operations inferior only to the Bank of England. Mr. Gilbart has announced in the ʻBankers' Magazine' for January last, that he will present the sum of £100 to the author of the best Essay which shall be written in reply to the following question :—

"In what way can any of the articles collected at the Industrial Exhibition of 1851 be rendered especially serviceable to the interests of practical Banking?"

And in the same announcement is contained an intimation of what such articles may consist, viz. :—

Architectural Models that may suggest improvements in the Bank House or Office.

Inventions by which Light, Heat, and Ventilation may be secured, so as to promote the health and comfort of Bank clerks.

Discoveries in the Fine Arts, by which the interior of the Bank may be decorated, or the Bank furniture rendered more commodious.

Improvements in Writing Paper, pens, ink, account books, scales, letter copying machines, or other instruments used in carrying on the business.

Improvements in Printing and Engraving, by which Banks may get their notes, receipts, letters of credit, and other documents, of a better kind, at a less expense, or so as to prevent forgery.

New inventions in the construction of locks, cash boxes, and *safes*, which shall render property more secure against fire or thieves.

And generally all articles, of every kind, which can be so applied as to *improve, cheapen, or facilitate* any of the practical operations of Banking.

This announcement appears to have a twofold object—first, to secure to a

B

large class of men a great advantage, by inducing them to give that amount of intelligent and careful attention to the scientific and really useful objects displayed at the Great Exhibition, which, without some such stimulus, would probably not have been elicited; and secondly, to promote the convenience and welfare of the Banking community, by pointing out, through the medium of a familiar and practical Essay, any objects which may seem calculated to assist its members in the prosecution of their business.

Although the proffered prize can be obtained by only *one*, the many unsuccessful competitors will have the consolation of not labouring in vain, inasmuch as they will have reaped a considerable share of practical and scientific knowledge, and have become the subjects of that wholesome mental influence which invariably accrues from the energetic direction of the mind to any object that widens its comprehensive faculties or dilates its powers. For this, at the least, every unsuccessful candidate will have to tender his sincerest thanks to the liberal originator of his Essay, unsuccessful though it be.

So much having been already most ably written upon the Great Industrial Exhibition, little apology, it is hoped, is needed for no further introduction of the subject. One suggestion, however, may perhaps not improperly be here introduced, in reference to the importance of a continued diminution of our fiscal burdens, by an extended reduction of our Inland Revenue and Customs, especially where those duties are imposed between the mother country and her colonies, thus practically aiding the object of the Royal Founder of the National Exhibition, which was to promote closer and more intimate relations between our own and foreign countries, by affording encouragement to international trade and reciprocity of commerce.

To this remark it may be added that, in the attentive study of the Exhibition, a subject for great thankfulness is found, in the striking parallel which, it has been observed, exists between the religious advancement of the several countries, and the state of science and manufacture there prevailing, as indicated by their productions, which they have respectively sent over, so prominently marked by our own position in the affair, establishing at once the fact that England's religious privilege and position stands intimately connected with England's productiveness and power.

As an Essay of practical utility is the liberal founder's object, and as objects of practical utility are the subject of the writer's Essay, the reader will, it is presumed, expect to find plainness and perspicuity of expression rather than studied elegance of style.

The announcement of Mr. Gilbart's prize having been accompanied by a very comprehensive framework for the Essay, it seems highly suitable to make *that* in some degree a guide as to the order of proceeding, and without confining the writer strictly to its limits, to class the objects of remark as much as may be under its several heads. It must, of course, be apparent that many, perhaps most, of the inventions and articles which will be referred to have not an *exclusive* bearing upon Banks or Banking, or upon the comfort and convenience of those engaged therein; but this exclusive adaptation is not required in the announcement, and it is therefore submitted, that the following observations are perfectly within its scope, if the objects referred to can be shown to possess advantages in which the Banking community may share, conjointly with other sections of the people generally.

The Essayist's attention is, first of all, directed to

I.—" ARCHITECTURAL MODELS THAT MAY SUGGEST IMPROVEMENTS IN THE BANK HOUSE OR OFFICE."

The architecture of a Banking House, more especially if it be of a National character, or expected to assume an important position from the magnitude of its operations, should be marked externally, internally, and everywhere, by *Stability*, as its leading feature; which a builder of intelligence will take care shall be combined with *Taste*.

In the present highly enlightened and civilized state of society in England, it may, at first sight, be thought by some unseemly to erect a Banking House on the principle adopted at the Bank of England, where the whole of the four external faces of the building present to the observer a regular series of fortification, having all its windows facing inwards, and accessible from the four streets with which it is surrounded only by massive gates and doors; but when it is remembered that the Bank of England is intended and expected (and what contemplated Banking House is *not* intended) to endure for centuries, and to witness possibly political convulsions, as well as the wear and waste of time, it will be seen to be wise, even in such peaceful times as these, to found it on a rock, and rear it like a rock—of granite; and supplied with means for the location and protection, in cases of emergency, of a Constabulary, or even Military Force, upon its roofs and ramparts.

The Ground Floor and the Basement will be objects of the first importance; and in digging for the latter, and in the construction of the areas, it will be well to prevent the possibility of the "*Screw*" being used in any attempt to force the shutters. This *has* been attempted where the areas are too much confined—the soil serving as a rest or fulcrum for the "*Double Screw*" to act upon—an instrument of prodigious power, before which almost any shutter must give way. For the same reason, Iron Palisading, when placed in front of windows, should be so constructed that, whilst calculated to resist pressure *from without*, it may not be needlessly strong in the opposite direction; otherwise it will only afford the means of using this powerful invading force, which is impotent when unsupported.

The principal features and more substantial parts of Banking House architecture, however, will generally be much affected by local circumstances, as well as by the character and amount of business to be transacted in the budding.

These, therefore, may, perhaps, be deemed—though in themselves considerations of paramount importance—yet, for the present purpose, as somewhat subordinate to those internal details and minor arrangements, upon which the comfort and convenience of the occupants of such buildings *as are at present existing* so materially depend; and which, it is presumed, the Proposer of the Essay primarily contemplated.

Amongst these may be suggested the character and style of *Windows;* the sort of *Shutters;* the arrangement of *Acoustic Tubes*, as well as *Stoves* or *Warming Apparatus, Wind Guards*, and *Cowls for Chimneys;*—not omitting, although more connected with the original construction of the building, *Fireproof Floors* and *Ceilings, Windows* and *Sashes.*

The External Doors should be of the very strongest character; and, if wood be used, it should be lined with iron, or thickly studded with rivets of the same material.

There seems also a strict propriety in protecting the doors at night (in accordance with the windows) by revolving Iron Shutters, of which more will be said hereafter.

The Inner Doors, forming the more immediate access to the Public Counter, it is submitted should not be, as at present, a pair of folding-doors, *one* adapted for entrance, and *the other* for departure; but should be two distinct and separate openings, as far removed from each other as circumstances will admit of; so that the approach to the counter should be at one end, and the exit at the other—the public always passing in the same direction.

Several improvements have presented themselves in Windows and Window Sashes; and "*Barrow's Patent Sash*," exhibited by Mr. Thomas Turner, is extremely good, and adapted for first-rate windows of plate-glass, which, since the recent abolition of the duty, has become considerably reduced in price. This sash is made in bronze metal, and the cost per foot is regulated by the size of the sash required. They are said to be extensively used, and that the Bristol and Manchester Branches of the Bank of England have adopted them. Each sash slides up and down, so that the window can be opened both at the top and bottom ; and the meeting bars are so constructed that the sashes when closed are perfectly flush on both sides, instead of the lower one, as is the case with common sashes, projecting its own thickness into the room ;—both sashes being in the same plane gives a far neater appearance to the window, and presents less lodgment for dirt or dust. It has, moreover, a self-acting latch in the sill-bar, so arranged that *the sash cannot be closed without being securely fastened.* The lines are concealed, and it is remarkably free from vibration.

The specimen exhibited was most beautifully executed, and would form a very suitable window for handsome buildings.

Mr. G. Hurwood, the inventor, exhibits a patent apparatus for "*Moving and Fastening Windows,*" the various applications of which are very ingenious. He has published an illustrated Pamphlet, containing a full description. One principal feature in the invention is the application of the screw, the worm of which, working into a rack or pinion, acts as a prime mover—dispensing with all pulleys, weights, cords, springs, set-opes, fastenings, &c.; and is particularly adapted to pivot windows, and to situations where the windows are above ordinary reach, as in lofty buildings. This invention has been very satisfactorily adopted in some of the public buildings in Norwich, and was availed of to regulate the ventilators in the Great Exhibition. It most effectually does away with the harsh and noisy friction of sash pulleys, which it is almost impossible to oil ; and overcomes, by the mechanical power of the screw, the difficulty frequently experienced in opening and shutting windows. The speed and power of the motion communicated can be readily regulated by the diameter of the rack or pinion employed to produce it.

Mr. George Anderson's model of an improved window has some excellent properties ; and he intends shortly to publish a full description, together with the prices.

Much inconvenience having arisen from dirty windows, and no small amount of danger having attended the cleaning of them, many have attempted to obviate the necessity for getting on the outer side for this purpose; among the most successful essays is, perhaps, that of Mr. C. Herring, who exhibits a patent sash, which can most easily be taken out to be cleaned. The arrangement is extremely simple, and consists of a metal rod dovetailing into the sides of the sash frame, and sliding up and down with the sash; this rod is connected with the sash by two small flush bolts at top and bottom of it ; the lines and weights are attached to this rod ; when the bolts are withdrawn the rods slide up and the sash is released, and may be removed from the frame. A similar model, upon the same general principle, is exhibited by Mr. B. E. Roberts, of Nelson Place, Clifton, who also claims the invention. As we shall have occasion to make a few additional remarks upon windows in alluding to the important question of ventilation, we pass on to notice

Shutters.

Shutters, as a means of protection, form an important feature in the fittings of a banking building, and, as some degree of solidity and strength, and consequently weight, is necessary, the arrangements for placing and removing them should be as scientific as possible, in order to economise both time and labour. For this purpose, the revolving iron shutters, now coming into general use, seem very desirable; by it, the antiquated plan of putting up and taking down, one by one, separate external shutters, is avoided, and the expense of replacing broken glass and repairing shutters, (ofttimes shattered by uncareful handling of them,) as well as the inconvenience, and sometimes accident, arising from the long bars by which they are fastened, is entirely removed.

The revolving shutters formed of curvilinear iron lath, and manufactured by Messrs. Bunnett & Co., Deptford, are very excellent in their construction, and work most easily. The curving of the iron lath, of which this shutter is composed, gives additional strength, without adding very materially to its weight, or much augmenting its expense. Their cost is usually from 6s. to 7s. per foot superficial, according to the size of the window, and, if kept well painted, are extremely durable.

These shutters seem only a modification of the principle adopted by Mr. Harcourt Quincey, in his Patent *Convex* Revolving Iron Shutters.

These are made of convex laths, and are said to be twelve times stronger than those composed of flat laths of the same substance.

These revolving shutters can be applied either horizontally or vertically, and are valuable, from the facility with which they may be closed in cases of popular tumult or of fire.

Mr. Archibald Horn is also the inventor of an Iron Shutter for Bankers; and in cases where it may be desirable to retain the old detached shutters, Jennings' Joints and *Shutter Shoes* are by no means without their use.

Acoustic Tubes.

Among the most useful means of saving unnecessary labour in offices may be placed the Speaking Tubes or Telegraphs and Communicators, by which messages may be sent from room to room ; and thus the necessity of running to and fro to answer bells, lessened by at least one-half. Mr. Whishaw's Telekerephona is a plain gutta percha tube, with a mouth-piece and whistle at each end; there are some additional arrangements for communications from several rooms; but the gutta percha pipe, mouth-pieces, and whistles, seem all that are necessary for ordinary purposes. Gutta percha is a most desirable substance for the tubing, as it possesses a remarkable sound-transmitting quality; and speaking-tubes, &c., are made by the Gutta Percha Company. The manner of calling attention to the pipe is by blowing into one end, when the whistle is immediately sounded at the other, and *attention* being first indicated, conversation may be commenced. When this is ended, the whistles, which are moveable, should be replaced. Messrs. Bryden and Sons have invented a *self-closing* mouth-piece, which would be a very desirable addition to these speaking tubes; by it, the removal of the whistle is rendered unnecessary. Perhaps these tubes might form a most useful means of communication between the Board Room and the several offices in the Bank, as well as from the counter to the Ledger Office, by which any inquiry might be immediately answered, without either loss of time or noticeable conference between the Cashier and the Ledger Clerk.

Wind-Guards and Cowls for Chimneys.

In few matters have there been so many different attempts to accomplish the same result as in the cure of Smoky Chimneys ; and it is very desirable that, if tolerated at all, the nuisance of smoke should be kept out of doors.

This subject is very intimately connected with ventilation, in the consideration of which it will be again referred to. One important means of obviating this nuisance is to prevent the free passage of the smoke, caused by the current of rarefied air up the chimney, being interfered with by the force of the wind striking the top of the chimney. To avoid this, Mr. J. E. Grisdale has invented a wind guard, consisting of a cap poised upon a point on the top of the chimney. This cap, the moment the wind strikes it, turns it back to the current, and closes that side of the opening. The principal objection appears to be in the extreme facility with which it vibrates. Mr. Grisdale has registered this invention provisionally, and, not having the means himself, is anxious to meet with some one to assist him in bringing it out. From its extreme simplicity it would be very inexpensive.

Mr. Stafford's interceptor cowl, for the same purpose as that last mentioned, is very well arranged, but would be more costly, on account of its construction.

These contrivances will· be very useful in situations where a chimney is overtopped by lofty buildings; but we imagine that in the majority of cases where inconvenience is experienced, the evil exists near the fireplace, or in the "throat" of the chimney, rather than at the top of the flue. This will be more fully exhibited in our reference to stoves and fireplaces.

Fireproof Floors, Ceilings, and Roofs of Buildings.

Specimens of this improved construction of fire-proof floors are exhibited by Messrs. Fox and Barrett; and one is certainly struck with the assertion, that the expense of these substantial and massive floors is not greater than those of ordinary timber. This is surprising, on account of the great weight which has to be sustained, the floor and ceiling being about nine inches thick, of solid material. The small cost may, however, be in some degree accounted for by the extreme lightness of the iron joists, the cheapness of the material employed in each successive layer, as well as from the very little amount of *skilled labour* required in their composition.

In cases of fire, the floors, composed chiefly of timber, afford, from their very construction, a supply of air to feed the flame; and, from the combustible character of the material, present very little hope of its progress being stayed. This invention will afford the means of diminishing the present extensive use of timber in buildings, by the almost universal adoption of which considerable danger is incurred. Mr. Braidwood, the Superintendent of the London Fire Engine Establishment, has stated, in evidence before a Committee of the House of Lords, that, by exposure for a few years to heat *not much above that of boiling water,* timber is brought into a condition somewhat resembling that of *spontaneous combustion;* and for the fire which occurred a few years since at Day and Martin's blacking manufactory no other cause could be assigned than the ignition of the wooden casing in which the hot-water pipes were enclosed. If these facts be correct, the character of the floors through which warming apparatus are generally made to pass must be a matter of importance. Considerable strength and solidity are given to the building generally by these floors; the iron bearing the *tensile* strain, and the concrete resisting the force of *compression.* A paper upon the subject, read at a meeting of the Society of Arts, has been published, which contains a full account of the arrangement.

This invention seems well adapted to those rooms in which books that have been filled and banking documents are kept; where security from fire is so necessary a desideratum, not very readily obtained above ground.

We now consider, secondly, those inventions by which

II.—" LIGHT," " HEAT," AND " VENTILATION,"

may be secured, and which so materially promote the health and comfort of bank clerks.

Light—Gas, Candles.

Like many of the discoveries of modern times, gas has not only great capacities for contributing to men's convenience, but also for their annoyance and injury, and this is most fully exemplified when this valuable substitute for the solar ray is employed excessively.

Most jets are adapted by their form and size to burn well a certain quantity of gas, and if less be turned on inconvenience is sure to arise, and if more is allowed to pass, incomplete combustion ensues, producing rapidly a most injurious and painfully oppressive condition of the air in the apartment. Burners are frequently somewhat inaccessible, or in the press of business they are disregarded until the smell of half-consumed gas is insupportable, and in order to regulate the light it may be necessary to adjust the cock several times in the course of an hour. To obviate this evil Mr. Biddell has invented a self-regulating gas burner which adapts itself to all variations of pressure, preventing the flame from either rising or falling above or below any desired height to which it may be adjusted, thus saving gas and preventing smoke. Its self-acting property depends upon the expansion and contraction of a compound rod in the centre of the flame, which, by means of a small lever and simple valve, shuts off or lets on the gas according as its pressure may be increased or diminished. This invention can be adapted to the *button burner*, which gives a whiter light; a light, however, which is more expensive than the red on account of its rapid consumption of gas, although the objection of cost would be perhaps overruled, when it is remembered that the whiter the light the less prejudicial it is to the eye, and the more complete the combustion of the gas the purer will be the atmosphere of the apartment. The introduction of pipes leading from the gas burner shades into the flue of the fire-place would very materially assist ventilation, and carry away the impurities generated by them.

The cost of this burner is about 7s. 6d., and they are issued regulated to a 2½ inch flame. Mr. Biddell has some very flattering testimonials from Messrs. Ransome & Co., Ipswich, the Gas Fitters' Mutual Association of London, and others, to the value of his invention, which appears especially adapted to the counters of Banking Establishments, as well as other places where the proper amount of flame from the burner cannot be made the subject of frequent attention.

Mr. Biddell's burner is furnished with a gauze covering at the bottom, which insures great steadiness of the flame by preventing any sudden draught from disturbing it. This addition of wire-gauze has also been used with some other burners recently introduced, amongst others with that of Mr. Winfield, which is excellent, and with the opal shades is becoming now very generally used, the only inconvenience of this improvement is the necessity of removing the chimney to light the gas, otherwise the gauze, *if fine*, while the chimney is cold frequently prevents the immediate admission of air necessary to support the combustion of the gas. The shade invented by Mr. Winfield merits attention on account of its reflecting qualities, its durability and cleanly surface, and the pleasant light which it diffuses. This will probably entirely supersede the copper-plate reflector, which, being quite opaque, has a tendency too much to darken the room.

It may be remarked, in passing, that gas burners are treated somewhat inconsiderately; they are lighted night after night for months together without any attention being paid to their condition, and thus, from some of the holes becoming partly closed, an inequality in flame is a necessary result. And then, in order to produce the same amount of light, the gas is turned on higher, and incomplete combustion, in some parts of the flame at least, is sure to ensue. Might it not be well to have an annual or half-yearly inspection of all gas burners in a Bank?

Candles.

Some of the inconveniences attending the use of candles may be considerably lessened by a little invention of the Rev. W. H. Jones—the "Acolyte," or Patent Safety Candle Cap; which cap being placed upon the top descends as the candle burns, and ensures regularity in the melting of the wax or tallow, thereby preventing the guttering and the dripping of the grease. *Price's Patent Candle Company* are the Licensees for its manufacture, and it is intended more especially to be used with their article denominated the " Composite."

It may not have been generally observed what an amount of smoke some candles produce, causing the same unpleasant smell, as well as painful effect upon the lungs, as that arising from the smoke of gas. The best Composite Candles, of which there were some beautiful specimens exhibited, smoke the least—the additional expense of which is indeed very little, and no doubt, if more attention were directed to the subject, even common candles might be materially improved.

Heat.

So large a portion of civilised society, and especially those employed in banks, being engaged in pursuits of a sedentary kind, the means of obtaining artificial heat in this variable climate are of much importance, and it is very interesting to observe the various improvements which have been recently made.

With all these, however, (and some of the fire places are exceedingly attractive,) the question of *stoves and open fires* does not seem quite settled in favour of the latter, and Mr. Edwards steps forward in defence of Dr. Arnott's invention, against the many who, in professedly copying the stove, have not adopted the principle upon which it acts, and thus brought both into disrepute.

The real Arnott stove is upon the principle of an extensive and moderately warm heating metal surface at about 200° Fahrenheit; it consists of a sheet iron box, which may be of any dimensions in proportion to the size of the room to be heated; it is divided by a partition into two chambers of unequal dimensions, which communicate freely at the top and bottom. A fire box composed of iron lined with fire-brick, rests at the bottom of the larger chamber; access is obtained to it by a door which must fit closely. The refuse of the fire falls into an ashpit below, here also is a valve for the supply of air to the fire box. The fumes and heat of the fire pass up and down, over and under the partition, giving warmth to the outer case; the smoke finally passes off by a flue into an adjoining chimney. The heat is regulated by the valve for admitting air; when this is opened widely, a large stream enters and combustion becomes active; when on the contrary the aperture is reduced, a comparatively small stream is admitted and combustion languishes. The temperature of the outer case is raised or depressed accordingly by the revolution of the heat and smoke round the division of the chambers, their power of giving forth warmth is expended as far as possible on the plates of the outer case, so as to be serviceable for the end in view, and it might be possible to exhaust the whole, for that end, by lengthening the flue or causing a greater extent of it to pass through the air of the room before entering the chimney. This apparatus certainly makes the most economical use of fuel of any species of contrivance for producing artificial heat yet known. Six pounds of Welsh coal or coke, of the value of one penny, will it is said supply an ordinary one for a day.

The prevalent fault in the Arnott's stoves generally manufactured, has been a diminution of the heating surface in proportion to the strength of the furnace.

All metal surfaces, however well the principle of a large superficies moderately warmed may be observed, raise the temperature by two means, namely, by radiation and by conduction. Radiated heat, which is that given by a common fire, is perfectly safe, but the heat produced by the air coming in contact with a warmed surface is injurious. The air which forms the medium' of heating the rest has been altered in its character, particularly in being desiccated or deprived of its humidity. It is necessary to counteract this result by artificial infusion of humidity into the atmosphere. Perhaps the best possible arrangement for this purpose is that consisting in a trough of water with a roller moving in it, and a similar roller forming a windlass about 2 feet above. Between the windlass and the roller an endless piece of towelling revolves. The bottom of the piece of towelling, passing of course through the water, it is only necessary to turn the windlass a few times in order to make the whole wet, and this process may be repeated as often as necessary. The vapour constantly arising from the cloth, will, if sufficient in quantity, make good the want of humidity in the stove-heated air.

It is rather singular that Mr. Edwards makes no allusion in his little pamphlet to this desiccation of the air by stoves, nor to the means of removing it.

In the stoves exhibited by Mr. Jeakes and Mr. Pierce, together with the great advantage of heating by *earthen* instead of *metal* surfaces, a principle is announced, which is but rarely used although it has been occasionally adopted for many years, viz., the introduction of the air from the outside of the building by a flue or pipe. This, upon a little consideration, will be shown to be very important, and it is possible that neither of these gentlemen has fully recognized the value of the principle which they have adopted ; we shall however refer to this again.

Gas has been asserted to be an economical heating agent—and Messrs. Feetham and Co. exhibit a very elegant stove of white porcelain. In many of the gas stoves, by other manufacturers, the deterioration of the air, and the smoke caused by combustion are entirely overlooked, and the stoves are made without any provision for a chimney at all.

The white porcelain will not however conduct the heat so rapidly as a surface *dark* and *rough*.

In the construction of some of the open fire-places, an effort has been made to bring the body of fire forward into the room as much as possible, without diverting the passage of the smoke. This object appears to be very happily obtained in a grate exhibited by Messrs. Dean and Co. which is the exact reverse of many, even recently constructed—the back is quite flat, the bars projecting outwards, like a section of an ellipsis. The surface of the fire is thus increased and the whole is placed in front of the *four* reflecting cheeks. This appears manifestly superior to the *round back* and *flat fronted* fire-places in common use. The execution of these specimens is very elaborate, and for ordinary purposes would be too expensive, the plain black being about £9 10s., and if with ash-pan and ornamental side £25, but the principle might be recognized in a much cheaper form.

It has been questioned above, whether the principle of introducing a fresh current of air to the grate and chimney, by communication with the external atmosphere has been fully appreciated, or its value rightly understood. Open fires have been objected to, for causing draughts, and closed stoves have been substituted.; air being introduced under them from without to ventilate the room, but the one great cause of these draughts as well as that of many smoky chimneys, and the frequent failure of attempts to carry off smoke from gas burners, by pipes, into the open air, has been altogether overlooked, viz.:—*mechanical exhaustion of the air in the apartment by excessive rarefaction in the chimney.*

For the purpose of drawing attention to the improvement adopted in Mr. Jeakes' and also Mr. Pierce's stove, to which the jury have awarded a medal and " special approbation," it may not be unsuitable, here to copy a paper which we brought to London on visiting the Exhibition, in order to submit to some manufacturers of stoves, supposing its suggestions to be wholly new.

For the well-being of fires generally, a constant draught or supply of air is necessary, not merely to support combustion, but also *to supply the partial vacuum at the bottom of the chimney,* caused by the ascent of the air, *it being highly rarefied by excessive heat.* The air in the chimney being thus rarefied and heated, *to a degree far above that in the room,* the atmosphere is exhausted, and in fact *pumped out of the apartment;* thus the pressure of the external air is much increased, and this is perceived by a strong current entering at every hole and crevice—when a door is opened the draught is extremely inconvenient, and has the effect of immediately reducing the temperature, causing, of course, much waste of heat. To obviate this is the object of the present suggestion, viz., to introduce the air necessary to support combustion in the grate, and to supply the rarefaction in the chimney, not from the front of the fire-place, as is now universal, (thereby drawing out of the room the air which has just been warmed,) but from the atmosphere outside the apartment, by means of a pipe or flue laid beneath the floor, and under the hearth-stone, and opening *behind the grate,* a space being left here, *carefully covered at the top,* so as to prevent any direct communication between the pipe beneath the floor and the flue of the chimney. In order to introduce air into the chimney, to carry up the smoke, without making so large a demand upon the warmed air in the apartment, which now under the present arrangement of open fire-places, as soon as heated in the grate, rushes wastefully up the chimney, the *front edge of the stove or fire-place opening, should be perforated or have a narrow space wholly or partially round it.* From these perforations or spaces at the back of the face of the stove, and between the outer edge of the face and the surrounding slips, would flow a constant current of cold air; this would constantly direct itself towards the fire and chimney, and thus become the upward current. The reason for having the perforation round the *front edge* of the fire-place opening is this, that the smoke from the fire may be *embraced* by the draught of air and carried up the chimney; if they were in any other position this might not be accomplished, and the smoke might be sometimes driven into the room.

A double set of grate bars under the fire would be very useful, one set to be withdrawn, and replaced by the other at intervals, when the bottom of the fire gets choked with dust; thus allowing the small ashes to fall through without the wasteful application of the poker or the shovel, which pulverizes the coal. This double set might work upon a rod like a hinge, and pass one through the other, thus shifting the support of the fire without at all disturbing its position, but merely removing the dust from under it. In reference to the general question of close stoves and open fires, perhaps for Banks the Arnott stoves would really be the more suitable. The circulation of the blood in some gentlemen is so much more rapid than in others, that in an office, there is rarely perfect unanimity in managing the fire; one opening the windows, whilst another puts on coals. The coldness of the office on winter mornings is objectionable, to say nothing of the half-hourly attention required by an open fire; whilst the Arnott stoves exhibited by Mr. Edwards, need replenishing but twice in 24 hours, and yet preserve a uniform temperature for months together.

Ventilation.

This subject is one of such vast importance, and yet one so generally and almost entirely disregarded, that considerable diffidence is felt in endeavouring to point out some of the facilities offered at the Great Exhibition for its accomplishment. Fresh air has been well described as " the food as well as the breath of life;" and to those who, owing to their inactive pursuits, *breathe less* than others, it is more important that the quality of the air should be most unexceptionable. Within the last two or three years only, has this subject been at all generally considered, although lately, numbers of interesting books and pamphlets have been published upon it. Ventilation seems to divide itself into two parts, quite distinct in themselves, yet bearing one upon the other, viz., the introduction of fresh air, and the carrying off of that which is exhausted, or foul. The latter has been long recognized, and in public buildings, to some extent provided for by openings in roofs, &c.; although in most Banking Establishments even this seems to have been too frequently overlooked. The former has been overlooked in the architecture of buildings generally, and neglected on account of the difficulty of accomplishing it without producing inconvenience. In summer, the evil is lessened, as windows and doors can then be opened with impunity, and the draught up the chimney is less. In winter, every precaution is taken to exclude the air by double windows and doors, and other appliances, and just in proportion to the success of these expedients, and the amount of fire kept up, is the rush of cold air, when a door or window is necessarily opened. Thus a change of rooms becomes essential, and when this cannot be obtained, or when at busy periods the Banking hours are much prolonged, great inconvenience is experienced from the vitiated air.

Perhaps at the present time there is not one room in a thousand provided with any means of continuous ventilation; indeed, with our present fire-places, this seems impossible; but with some other plan for supplying the chimney draught, it is presumed we might with great comfort employ Lockhead's perforated glass, manufactured by Swinburn & Co., for some of the window squares; or Naylor's new glass ventilator, very suitable for a window, and almost frictionless. Moore's patent lever ventilator is admirably adapted to the same purpose.

Other kinds are designed to be fixed in the wall. The little ventilating window by Hart & Sons is deserving of notice; it has a self-acting safety bar to prevent its being opened or shut violently by the wind.

The ventilating bricks and windows manufactured by Warner & Sons are made ready to be built into a wall, and the frames being of iron, form with the glass most durable ventilators.

The specimen exhibited by Mr. J. H. Boobyer, of Harrison's registered Venetian ventilator, is extremely ornamental, and well suited to a papered or a coloured wall. The ventilating facilities afforded by the use of Ridgeway's hollow bricks must not be overlooked; the wall being thus formed of a complete series of air passages, a communication may readily be opened in any part with the external atmosphere; and for this purpose some little iron valves are used in Prince Albert's model cottages. Sherringham's ventilators, manufactured by Hayward, Brothers, would be applicable to this purpose; although it may be questioned whether the air will be found to take the precise direction indicated.

But one other ventilator demands attention, viz., Dr. Arnott's Chimney-valve, which is really valuable. All attempts to ventilate rooms as at present constructed, by the occasional opening of a door or window, are insufficient, the warm air being lighter than the cold. Just as a bottle of oil inverted in a stream of water remains full, because the oil is lighter than the water, so the part of the room above the level of the chimney opening remains full of a poisonous gas, because it is lighter than the current of pure air which passes from the door to the fireplace; and it will be remembered that the portion of the air which is breathed by all but children is *that part above the fire-place.*

For the complete working of these ventilators, which are manufactured in great variety by Mr. Edwards, it is obvious that an increased supply of air is necessary; this it is proposed to bring from a valve in the door, which may be very well, when the exhaustion by the chimney is otherwise supplied; but at present some susceptible folks might well object to sit midway in such a draught, if there be any truth in the Chinese proverb—" Avoid a current of cold air from a narrow passage as you would the bite of a serpent." It should be borne in mind that every man requires *four cubic feet of fresh air every minute*, or a quantity twice his own size, and that a common candle destroys as much, and gas burners much more; and we trust that such a happy combination of ventilating expedients may be adopted in banks generally, that fresh air may be obtained, and no one inconvenienced.

Health and Comfort of Bank Clerks.

It has been found to be very necessary to change the air of rooms as frequently as may be, and upon the same grounds it might be shown to be desirable to avoid all sources of unnecessary contamination; one very frequent source is the smell arising from drains, cesspools, sinks, &c. Some very efficient Effluvia Traps were exhibited. That made by Messrs. Bunnett is self-acting and self-cleansing, and with a fine grating, would act very well in situations where there is a considerable flow of water, and where the fall of the drain is sufficient to carry everything away. The Trap made by Messrs. Wilson and Woodfin is admirably adapted to cases where some amount of sand or sediment has to be removed, and where the fall of the drain is not sufficient to carry it away. For the same situations, and especially for sinks; and also for lavatories, Lowe's Patent Effluvia Trap Gratings, or Stench Trap Grids, are very useful; by these the usual " *syphon bend*" in the pipes from hand-basins, which is adopted as a cheap means of forming a trap, might well be dispensed with; thus avoiding the evil arising from sand being washed down the basin and lodging immoveably in this bend, whereby the pipe is obstructed.

Clark's Self-acting Valve Trap is also very good, and exceedingly simple in its construction, being specially adapted for sinks, &c.

This very comprehensive division of " Health and Comfort of Bank Clerks" seems to invite a suggestion as to whether it might not be desirable in all establishments where the persons employed are numerous (and why not in Banks?), to have fitted up, for occasional use, a warm bath apparatus. This may cause some to smile, but, upon a little consideration, it may appear by no means unreasonable. The desirability of such an indulgence occasionally will not be disputed. Then it will be remembered that a large proportion of bank clerks are in lodgings, that the expense of warm bath furniture is very considerable, and that, practically, very few houses are so supplied. Also, that the occasions when a warm bath is most useful, are those of slight cold, or some other little ailment, which requires attention, although a diligent and active clerk would not wish to be detained at home by such complaints. In such cases, perhaps, a cold would be increased by walking to a distance, and returning afterwards, and the expense of a cab added to that of the bath, would probably amount to four or five shillings, which, in many cases, would be prohibitory.

The bath would be but an adjunct to the usual lavatory, and the same room might suffice for both. The heating apparatus is so simple that very little contrivance is necessary for that. It may be connected with the warming arrangements for the Bank, or a little stove may be applied purposely, as in some of the baths referred to below.

The most delightful bath at the Exhibition is made by Mr. Finch, lined with porcelain tiles, the edges and corners of the bath being all turned and rounded, presenting a surface *necessitating* cleanliness.

Mr. Benham's copper bath is beautifully fitted up, with polished mahogany case, and combines facilities for hot, cold, shower, and vapour bath, price

£23; but it may be questioned how far *polished mahogany* is suitable for the purpose.

Those made by Messrs. R. & W. Wilson are intended to save the cost of a wood frame, in which all baths with taps and levers for hot and cold supply, and waste-pipes, have hitherto been fitted, whilst at the same time an ornamental character is given them; indeed, the ten different specimens exhibited are well deserving of attention. They are made of the best "Charcoal Iron Plates" tinned by themselves. The most suitable for the purpose alluded to would be the "Grecian Bath, with Taps and Levers," square, white inside, and imitation Siena marble outside.

Where no facility exists for obtaining hot water, the baths, combining heating apparatus with their construction, may be employed. Messrs. Tyler & Son have given much attention to the construction of bath apparatus, and have published a little pamphlet of their various contrivances for heating. Their "Villa" warm bath, with small heating stove, is very compact; price £15 15s., subject to 5 per cent. discount, and is completed for connection with the cold and waste-water pipes. Mr. R. Dale also exhibits an improved warm bath and heating apparatus, and Messrs. Warner & Sons the same.

The lavatories in the retiring rooms at the Exhibition have been so much appreciated that it is almost needless to call attention to them, except to point out the superiority of those having the back and horizontal slabs, to prevent the water from splashing the walls, &c. These are termed "Complete Fountain Hand-basins."

The hand-basins exhibited by Mr. Jennings have a very elegant appearance. The cocks are self-closing and exceedingly simple, and the waste-cock prevents the possibility of impure air rising. They work below, so that nothing is visible but the engraved "pulls." The water rises from the bottom of the basin like a fountain. In connection with these the glass pipes with patent joints, manufactured by Messrs. Swinburn, may be noticed, as very suitable for *rain water*, with which hand-basins should be supplied, and for the conveyance of which *lead* is unsuitable, the rain water, on account of its *comparative purity*, acting more powerfully on the lead than that from springs or rivers, and causing a poisonous quality in the water, which, if accidentally drank, might be attended with serious injury.

There are worthy of notice one or two .

III.—" DISCOVERIES IN THE FINE ARTS, BY WHICH THE INTERIOR OF A BANK MAY BE DECORATED."

Few persons can have passed round the Coloured Glass Gallery without being struck with the imposing decorations, consisting of full-sized representations of Distinguished Individuals, by Mr. Norwood. The figures apparently in carved oak upon a crimson ground; the skirting and cornice being quite consistent with the rest; so that it is difficult to realize the fact of its being a perfectly plane surface. These figures are printed from blocks, at a very cheap rate. Mr. Norwood has eleven figures, viz:—Her Majesty the Queen, H.R.H. Prince Albert, Lord Nelson, Sir Robert Peel, Shakspeare, Milton, Lord Byron, Robert Burns, Sir Walter Scott, Ceres and Mercury—emblems of Agriculture and Commerce. These, although the execution is by no means rude or coarse, are printed at 3s. 6d. each; and it would seem that if arrangements could be made with Mr. Norwood for the representation, in this way, of some eminent and distinguished Bankers, such portraits might form a very suitable decoration for the Board Room; and thus, at a trifling cost, perpetuate the memory of those who ought not to be forgotten.

Horn's Carved Oak Decorations, with knotted oak panels, are magnificent, and are an excellent imitation of polished oak of the finest grain. These can be had of any tint or colour; and are said to be extremely durable,—the varnished surface admitting of the wall being cleaned. Price, from 21s. to 30s. per piece.

As an interior decoration, the imitation marbles, painted upon wood and slate, might be used. The colouring surface upon the slate seems very firm; and the specimens exhibited by Mr. Hopkins, Mr. Smith, and Mr. T. Stirling, jun., are especially worthy of notice, on account of their beautiful resemblance and fineness of grain.

A variety of circumstances have lately conspired to make a general and comprehensive knowledge of geography extremely necessary, if not in the daily routine, at least in the occasional varieties of a Banking business; and *Maps*, used as decorations of a room, tend greatly to facilitate its acquisition.

Their size and their selection would, of course, be dictated by the situation, and by taste, and other circumstances. If intended to hang constantly, they should be well varnished; and Mr. Manning's varnish, made without the aid of heat, looks very well upon the Maps exhibited.

For embellishments generally, and for ornamenting ceilings, the Patent Earthenware, for which Mr. Hayes is agent, is recommended by its imperishable and cleanly character—requiring no renewal; and when fixed, not likely to get chipped and broken, like plaster. It admits of a very high degree of ornament, and in colour is perfectly unchangeable. The specimens, in sharpness of outline, appeared equal to any wood-carving or plaster impressions, if not to gutta percha.

This same material, in a plainer character, as also glass, seems well adapted for external use;—for lettering, for bell-pulls, and a variety of small matters, for which brass, &c., are at present used,—necessitating daily labour in polishing their surface.

Some of the *Letters* exhibited by Mr. Lee are really elegant.

IV.—DISCOVERIES BY WHICH THE BANK FURNITURE MAY BE RENDERED MORE COMMODIOUS.

This subject comprehends much of the minute; which, however, should not pass altogether unnoticed. The first object which presented itself was an Alarum Bedstead—suggesting to Bank officials the idea of *punctuality* in the morning, and the thought may not be altogether worthless for the winter of 1851.

Door Balances.

Mr. John McClure has invented a very great improvement upon our common door springs.

These are usually weakest when the door is shut, and stiffest when the door is opened wide; just the very reverse of what they should be. "The balances" are heaviest when the door is shut, and quite light when open. These are not in the market yet, having only recently been registered; but, from the models exhibited, they bid fair to supersede the springs. The A balance, with double connexion, for bank-doors, would be the best: they can be applied in various positions; the best place would be at the bottom hinge. The apparatus will be contained in a cast-iron box, about 9 inches square, and 5 inches deep, which is to be sunk into the floor; and the balance-weight requires a depth of from 9 to 11 inches to work in.

For ordinary light doors, Scott's *Silent Spring*, of vulcanized india-rubber, may be useful, on account of its cheapness, viz., 2s.

The passing from one room to another may be much facilitated by Windle and Blyth's

Registered Bevil Lift-latch,

which, by simply pulling the knob towards you on one side, or pushing it from you on the other, opens the door; the words "Pull," "Push," being engraved on little ivory plates on either side.

Self-acting Door Fastenings.

In a retired corner of Class 22 was placed the model of a door fitted with improved fastenings, made by Mr. Greenfield expressly for the Great Exhibition. These fastenings are modifications of those at present in use, with some slight additions. He has arrived rather scientifically at the conclusion, that his latch requires only one-eighth of the force usually necessary to close a door; and asserts, that these fastenings are not liable to get out of repair for many years, however rusty. The door is made to close without noise or vibration, and cannot be left ajar. When the fastenings are attached for the night, the shutting of the door locks them, and they then require two hands to operate at *different parts* to open them—this being a security against the plan of cutting a hole through the door to shoot the bolts. It is to be regretted that these fastenings cannot possibly be *described* intelligibly without an illustration; but they will well repay the trouble of inspection. Mr. Greenfield offered to put the additional fittings to a lock and two bolts adapted to an ordinary size front door for 5s. The principal value seems to be *additional security* without *additional trouble*.

Floor Cloth.

For quietness and durability the " Kamptulicon Floor Cloth," manufactured by Walter & Gough, is peculiarly distinguished. The price, however, is very high.

Scraper and Foot Brush.

That which tends to promote cleanliness in an office where a number of gentlemen pass so large a portion of their lives, cannot be called trifling or useless. The "Improved Hall Foot Brush," registered by Thompson & Co., is of this class; and it may be well to notice here the importance of daily *sprinkling the floors before sweeping*, to avoid the needless accumulation of dust upon book-shelves, papers, books, &c., whence it is very unpleasantly transferred to the fingers of those who handle them. There is a prejudice against this method, because it stains the floor; but better stain the floor than spoil the books and soil one's hands.

The " Hand Pressure Table Bells'

should also be noticed, as being very suitable for a manager's room, manufactured by *Simcox, Pemberton, & Sons;* as may also the *German Thermometers*, to be had of Mr. Pritchard, Fleet-street, for which our continental neighbours are so famous.

Harcourt Quincey's

" Patent Pedestal Coal Vase"

is deserving of attention—saving the trouble of lifting a heavy coal scuttle, and the occasional inconvenience of an empty one. The

" Folding Library Steps,"

made by Mr. Ell, would be found very serviceable, on account of their lightness, strength, and firm standing : well adapted for Bank rooms with high shelving. The sides are bowed, and the steps, which are dove-tailed into them, have risers tongued to their backs. These make a thin, light piece of wood, as strong as a thicker and heavier piece would be without them. Those at the Exhibition, French-polished, with lacquered hinges, price £3 10s.; same in American ash, £2 5s.; and in deal, £1 15s.; and if made higher, proportionately more.

Fire Annihilator.

Phillips' Patent Fire Annihilator, unless its power is greatly exaggerated, would form a very valuable adjunct to the furniture of Banks, where the destruction of books might be irreparable. Its great advantage seems to lie in immediate readiness for action, thereby affording the opportunity of checking a fire at its commencement.

Venetian Blind.

The New Venetian Blind, made by Hopkins and Son, is a great improvement on those in ordinary use. It is drawn up and lowered by one single cord. By drawing the cord a little *"from"* the window the blind drops to any position you may require, and becomes fixed by bringing the cord again into a *perpendicular* position ; thus dispensing with the knots and hooks necessary hitherto.

For an ornamental table or slab, the

Imitation Marble on Glass,

exhibited in Class 24, Nos. 37, 38, 39, and 40, appeared very suitable, and would, probably, be cheaper than real marble. The Siena sample, No. 37, is an excellent imitation.

Bank Counter and Money Drawers.

There is in the new office of the National Bank of Scotland, Glasgow, a beautiful counter, of novel construction, having a sort of demarcation line between that portion appropriated to the customer, and that retained exclusively by the Bank Officials ; and in connexion with the various specimens of glass exhibited in the Crystal Palace, the idea occurred, whether the friction, in telling coin upon wood, might not be diminished by substituting some mineral, or metallic surface ; possibly of glass, which would admit of being easily cleaned. The quality of the "ring" produced would be of much importance, and this could only be discovered by experiment. But whether glass counters would do or not, there can be little doubt that *moulded glass,* or *glazed earthenware money boxes* for counter drawers, would answer well, and might be made much cheaper than the elliptical mahogany boxes, turned by an eccentric lathe.

Safety Collecting Boxes.

In the press of business, a place of safety, immediately accessible, is necessary, for bags of coin, paid in by customers, and set aside to be told at leisure. The Railway Collecting Boxes, made by Mr. Fisher, seem applicable to this purpose ; the bags being put into the top, the shutting of the lid causes an open cylinder to revolve, and the bag falls to the bottom ; it cannot then be seen or touched from the top opening, but can be removed by unlocking a door in the side. It will be observed that small "bags" of coin offer such peculiar facilities for appropriation, that they should not be unnecessarily exposed.

Fixture for Tin Boxes.

The daring robbery at the London and Westminster Bank, in June last, suggests the necessity of some ready method of securing the tin boxes in which specie, notes, bills, &c., may be kept, in the immediate vicinity of the counter. These boxes might have some slight projection behind, fitted to a corresponding recess in the back of the ledge or shelf, on which they stand. Several, being thus placed in a row, may be secured by a rising and falling iron bar in front, locked at one end.

Specie Travelling Trunk.

For the conveyance of specie from place to place, nothing, perhaps, unites ready portability and safety, so well as a small thick leather or gutta percha trunk. That made by Mr. Motte, out of *one piece*, would seem suitable for this purpose.

The American riveted portmanteaus, Canada Department, No. 196, in which no thread or sewing is employed, obtain great strength and durability by the rivet heads projecting slightly from the surface; and for travelling with papers, &c., for the inspection of branches, or otherwise, the commodions bag with improved fittings, exhibited by Mr. Meller, would combine many conveniences. It is lined throughout with best Morocco leather, has metal knobs at bottom, to prevent wear, improved secret fastenings, four pockets, &c.—Price 5½ guineas. This is of course made in the very first style. They may be had of *enamelled* leather, and lined after the same design, for two guineas.

V.—IMPROVEMENTS IN WRITING PAPER, PENS, INK, ACCOUNT BOOKS, SCALES, LETTER COPYING MACHINES, OR OTHER INSTRUMENTS USED IN CARRYING ON THE BUSINESS.

Improvement in Paper.

The quantity of writing paper used in banking business is such, that the price at which it can be obtained is of considerable importance; and this, together with the quality, would greatly determine as to its merit. The specimen books of writing paper by Mr. Ralph, are deserving of attention, as uniting cheapness with good quality. Some improvements have been exhibited in *peculiar kinds* of paper, by which additional security may be obtained against fraud, and for the transmission of cash, &c. There are also some inventions for preventing fraud, which will be pointed out under the head " *Engraving ;* " but *Saunders' Watermark* claims attention here. Hitherto (with the striking exception of the Bank of England and a few others) bankers have sought to protect themselves from the imitation of their notes solely by expensive, elaborate, and extremely difficult engraving; and although the Bank of England protects HERSELF by certain secret and almost invisible marks, yet the public at large are constantly exposed to danger from taking bank notes which, for want of these secret marks, are afterwards detected to be forged. The advantage, therefore, of obtaining additional protection from a *watermark* in the paper, difficult or impossible of imitation (except by a very few manufacturers), is manifest. The expense of a *variety* of elaborate watermark designs for different bankers is not necessary, it not being so much a matter of taste as of bare utility; and ONE very complete and rare design might serve, in the manufacturer's hands, for. *bankers' note paper* generally. This might then become a sort of guarantee with which the public would be acquainted, and would prevent the palming off of forged notes upon individuals who cannot at present judge, at a glance, of their genuineness. An additional security would be thus obtained at a very small additional expense, and one, in the benefits of which the public would participate; and notes, instead of going out after breakfast and coming back to dinner, might not go home till the morning. The process has been patented by Mr. Saunders, and its principal feature is, that designs in outline, *and in shade,* can be introduced, as watermarks into paper, they having hitherto been confined to words in Roman characters, or devices of the simplest kind, and in outline only.

Mr. Wilde's " Floreated Writing Paper" is also a very good specimen of water-mark, and although a different process to that of Mr. Saunders, for all purposes of mere *outline,* seems sufficient. It is remarkable that this unequal thickness, by which we suppose the water-mark is produced, should be imperceptible to the pen; practically, however, the surface of the paper cannot be *improved* by it for writing.

C

Check Paper.

For business purposes, although not quite so handsome as an elaborately engraven check, printed in vegetable colours, the paper made by Nissen and Parker from chemically prepared pulp, may answer well, especially as it immediately betrays the action of chemical agents. Thirty-two specimens of chemical application to this paper are exhibited, the effects of all being sufficiently obvious. The colour of these checks renders the tracing of a signature impossible, two checks laid one upon another being quite opaque. The process is cheaper also; and these checks, although not so ornamental, have an advantage over some heavily engraved, which, if written on with pale ink, are hardly legible. The difference, of almost 20 per cent., between the cost of checks from the plate and those printed from type, seems rather considerable, and if some uncommon type were used, perhaps the security would be as great as in using engraved checks. Moreover, cases of forgery by an imitation of the check are very rare, the facilities for obtaining genuine forms being far too great; and until customers will co-operate with their bankers to prevent fraud, by rendering their check-books less accessible, the inimitable character of the check is of comparatively small importance.

Metallic Paper.

The indelible character of style writing, upon Penny's improved Metallic Paper, may recommend it for some purposes where the use of ink is inconvenient; and being nearly, if not quite, as permanent, it might be useful for exchange and clearing books. The paper exhibited was remarkably easy to write upon, the unpleasant friction of the style being entirely absent.

Brown Paper.

The glazed brown paper exhibited by Venables, Wilson, & Tyler, has some valuable properties. Its smoothness renders it very pleasant in use, and it is *firm, strong*, and *light :* this latter quality is of much importance; for the price of brown paper being very much regulated by its weight, the thinner the paper, compatible with the strength required, the more advantageous to the purchaser. This sample is fifty inches wide, and may be had in any length. It would be well suited to £5 silver parcels, and banking purposes in general.

Mr. Martin has invented a non-absorbent paper size, which he has patented. Instructions for its application can be obtained, at a moderate charge; and it might be very usefully employed for the brown paper used for Bankers' parcels, which are often endangered by exposure to wet. In the specimens exhibited of prepared and unprepared Blotting Paper, the difference was most remarkable;—the one instantly absorbing the water placed upon it, and the other throwing it off immediately.

Blotting Paper.

This being an article of large consumption in Banking business, is of much importance. Its present condition seems capable of much improvement, and would well repay a little attention. The blotting rollers are ineffectual, and the *blotting sheet* and hand seem the only alternative. Blotting paper seems generally to be good in proportion to its rottenness, and *vice versa*. It, therefore, is important to lessen the friction of the hand upon it, which progressively injures its absorbent quality, and causes it to tear, before half saturated.

It would afford much facility, if some glazed surface could be readily attached to a fold of blotting paper, of some twelve or sixteen thicknesses. These might be torn off as used.

Envelopes.

Notwithstanding the well-founded objections which have been raised against the use of Envelopes, they are generally adopted; and, on account of their convenience, are especially adapted for enclosures. Much time is also saved in using them, as unnecessary folding is avoided. The newly-invented *Paper Cloth* is well suited to parcels containing money or valuable documents.

Stone's Patent Bankers' Safety or Parchment Paper, manufactured by Mr. Saunders, may answer the same purpose, although the expense would be too great for ordinary use (£7 10s. the ream of 500 sheets). It is light, and very strong. Many persons must have observed the sheet at the Great Exhibition, sustaining a weight of 4½ cwt., and attached to which was the following certificate :—" We hereby certify that a sheet of paper, of the same substance as the one exhibited, weighing less than 1¼ oz., sustained, without fracture, 5 cwt. and 24 lbs., being the utmost weight we could attach to the apparatus ! (Signed) WM. WILLIAMS and SONS, Scale Makers, Cannon Street, City. 19th April, 1851."

This is a very remarkable paper; and when wet, it will stretch considerably before breaking. The process might, with advantage, be applied to Bank Notes, and similar documents, which have to endure great wear.

The Envelope Machines of De la Rue and Co., and of Waterlow and Sons, have deservedly received much approbation, and to these and other improvements is to be attributed the striking reduction in the price of envelopes within the last few years. The machines have been so well described already, that a repetition is unnecessary; by a very slight difference, however, in the dies of the two machines, one important variation in the form of the envelopes, is produced. In those by De la Rue and Co., the ends are first folded down, then the bottom and top flaps. In Waterlow's envelope, the bottom flap is first folded, then those at the ends, and then all are covered by the top flap. This is, obviously, the most useful envelope for business purposes, and especially for enclosures, as the *seal catches all four flaps;* whereas in all envelopes (and they are the greater number) made like De la Rue's, the seal very frequently, only connects the top and bottom folds, in which case, the gum being first softened, by steam or moisture, the ends may be drawn out, the remittance abstracted, and the envelope restored to its original appearance. This has been done in the specimen, although the cement seems very good.

The Polychrest Envelope, invented by Mr. Ralph, is well adapted for security. The bottom flap is cut long, and when folded, it reaches beyond the opposite side of the envelope; it is then turned back, and thus embraces the two ends, and lastly, the top fold covers all closely. This envelope, however, might be improved by cutting away the end flaps, a little at the bottom, so as to allow the seal to hold all four: the price for large letter size is, 15s. per 1000.

For the transmission of " *Crossed Checks*," the adhesive envelopes save much trouble in sealing, and there seems no reason why *post paper* might not be prepared in the same manner, for ordinary letters.

Great facilities have recently been gained by the use of envelopes *with printed directions*, for transmitting to other bankers, checks drawn upon their firms, in the benefit of which the post office has participated, in the facility with which such directions are read. It would be well if, in return, an opportunity were afforded of obtaining POSTAGE ENVELOPES at a fairly remunerating price.

Pens.

The *Quill* has become practically superseded by the steel pen, and unless some means can be devised of rendering ink and steel pens less antagonistic, the latter, in turn, must be displaced by something better adapted to the purpose. For many years the price of a gold pen was £1 1s.; recently, indeed, by the competition amongst retail dealers, a Mordan's pen may be got for 19s.; but this has been found too much, and a cheaper, although effectually anti-corrosive pen, has been extensively introduced, and now gold pens, although not so good as the £1 1s. pen, may be got for one-fourth of the money. Some of these are made by the first makers, and it is very desirable that in a thing of this kind, so much depending upon the care bestowed upon its manufacture, the maker's name should be permitted to appear. If the price is not first-class, no one would expect a guaranty of A 1 excellence. These pens, although little influenced by ink or friction, are not indestructible by the pressure of ledgers or the opening of desks, or falling pointedly upon the floor. Some means are requisite for their protection. These may consist either of some kind of *rest* m which to place the pen for a moment, when laid aside, or of a holder, such as that exhibited by Mr. Mallett, which draws in, like a pencil case, and which might be further improved by being slightly weighted at the end farthest from the pen, that in falling the nibs might not be first to come in contact with the ground. The *rest*, however, would be far preferable, as in some departments of book-keeping, the pen is perpetually laid down and taken up, and the only safe position for it, is between the lips, where the bitter varnish immediately reminds one of its impropriety. Indeed the unpolished cedar is better for pen-holders; its absorbent quality rendering it more pleasant to handle in warm weather.

Mr. Mallett's gold pens are made to screw into the holder, and are provided with a little fork underneath, to prevent the nibs from crossing. Prices, 6s. 6d. for the "platina pens," and 13s. for those with "ruby points." The little fork would somewhat interfere with the cleaning of the pen, for which gold offers such delightful facilities, and which should be further assisted *by both sides of the gold pens being polished.*

Mr. Myers exhibits some excellent gold pens, in *quill* holders, which, on account of their lightness, are very pleasant to use, and, excepting the *halfpenny cedar holder*, perhaps the best. His steel pens are also very good. The splendid display of steel pens by Mr. Gillott was sufficiently striking. It seems a pity that such elaborate and excellent workmanship should be dipped into ink, only to be presently destroyed; for, notwithstanding such vaunted excellence of many pens, including those of *Albata* metal, they *do corrode*, and that in *any* ink, requiring to be continually scraped upon the "*coupon*" of a cheque book, or something else, to keep them in usable condition. The *nibs* being so soon spoiled, the use of *barrel pens*, which are much more expensive, seems undesirable. The idea of a bank providing all its clerks, once in two or three years, with a gold pen each, sounds rather extravagant, but it might not cost more in the end; and Mordan's pencils, at 6s. a dozen, with black-lead and soft cedar, are really cheaper, if used with care, than others at 2s.

Ink.

In consequence of the general use now of steel pens, ink requires to be made much thinner than formerly, and this very much lessens its durability. This will be very obvious on looking through the Court Rolls of any manor, or other ancient manuscripts, extending consecutively over a long series of years: and the evil is still increased in the case of *letters*, which, by the copying machine process, are robbed of half the ink put upon them.

This failure and deterioration of ink is a very serious evil, and, it is feared, must arise either from the absence of some important ingredient, dispensed with, perhaps, from a penny-wise economy, to which the manufacturer is goaded by the competition of his brethren, or else from the contact of *steel* pens in writing.

The truth appears to be, that whereas ink used to be composed of *gall nuts*, the object is now, amongst ink manufacturers, by chemical appliances, to produce a black fluid at the *least possible expense.*

If, upon consideration, it seems desirable that more attention should be paid to the inks which are used in banks, might it not be a good arrangement to purchase the best ink of the first maker (its carriage being very trifling), giving him notice, that a memorandum of its purchase would he made in the stationery account, together with a note, that all the books in the office would be written with it, thus furnishing a lasting record of its durability.

We have in our possession an ancient deed 506 years old, and yet the ink shows but slightly any deterioration in its colours; whereas some of the modern ink, sold by respectable manufacturers, and used but six years since, looks far less black and bright. From the great importance of the durability of ink, it is desirable that much attention and long experience should be brought to bear upon its manufacture; and to remunerate this care, the best ink should not be considered dear at the price of the best of wine.

Although desirable to have all inks as fluid as may be, yet for copying some degree of " body" is required; and this, for the sake of durability, should not consist mainly of glutinous material.

Mr. Todd has made it his business to manufacture ink scientifically; and the *" Perth Writing Inks"* are justly celebrated. Some writing in this ink, dated 1830, is shown. This, although in very fair condition, is not at all equal to those of olden time.

The copies produced from the Perth Copying Ink, point it out as desirable for this purpose, being very black and legible. The distinctive feature of the ink manufactured by Mr. Lovejoy, is its capability of resisting the influence of the ordinary *chemical re-agents*, by which other inks are obliterated. In the samples exhibited, the acids seem to have given *additional strength* and permanence to the writing, although in some cases the texture of the paper itself was destroyed. There were specimens tried by nitric, muriatic, and sulphuric acid, fourteen times diluted; by iodide of potass, one part to six of water; by citric acid, or salt of lemons, one part to four; and by liquor potassæ and oxalic acid, one part to three of water. This ink would be very valuable for writing cheques, bills of exchange, and letters of credit, to alter the *amount* of which there may be a temptation; and a double security would be obtained, by using the indestructible ink upon the sensitive tinted paper.

Ink Bottles.

The Ink Bottles, with a lip for conveniently pouring, made by Mr. Isaac, are a considerable improvement upon the old stone bottles, from which it is almost impossible to take ink cleanly.

Account Books.—Binding.

The same remark which was made in reference to writing-paper applies also to account books—that *consistent economy* should be studied as well as *superior quality;* and it is quite as true that some books are made at a needless expense, as that others are not nearly so substantial and strong as they should be. Some of the subsidiary banking books are practically done with when once filled, and that will be sometimes in six months or less. Others, such as current account ledgers, are *never* done with; and although, when filled, sometimes in one year, sometimes in five or six, they may not have to sustain much heavy wear and tear, they are frequently referred to. A third class, such as character and personal record books, may be in use for twenty years, or even longer periods; the renewal of these is attended with great and protracted labour. It is manifest, therefore, that the style of binding for these several kinds may widely differ. The first class should be bound as cheaply as may be, consistently with comfort, during the short period they are in use. The second class should be very strongly bound, considering not merely the time they are in *daily* use, but the indefinite period during which they may be required for reference; and for the third class, no binding can be too good, as far as materials and strength may be concerned;—the giving way of the binding sometimes rendering weeks of labour necessary, in transcribing, or the perpetual annoyance of a disjointed book. These remarks were suggested by the samples of Account Books exhibited by Messrs. Sibel and Mott; one set being made to order, for the City Bank, New York. The binding of these is super-excellent; and the ruling most elaborate and beautifully executed. The marginal lines, consisting of four colours, mitred at the corner, and the books are gilt-edged; the folio being put in the *middle!* of the page, instead of the corner.

The trade account books were bound in very handsome style indeed, and for such binding, paper, ruling, &c., the price may be considered very low. One set foolscap, 3s. per quire; one set demy, paged, 8s. per quire; and one set medium, paged, 9s. per quire.

The account books exhibited by Mr. Möller are deserving of especial notice, both on account of their character and price. Three books were exhibited, two of which were of practical utility, although rather "ultra" in their style of ornament. They opened remarkably well, and the indices, being only half width, were fixed with brass hinges to the middle of the cover, so as to be more readily referred to. One of them, with extra ruling and printed head, 9 quires demy, very good paper, and strongly bound in "Jucht" skin imported from Russia, price £2 (see Hamburg Priced Catalogue, page 3.) The other, very handsomely bound in "Jucht" skin, with Russia bands, and of thicker paper, ruled, and with printed head and gilt edges! price £2 10s. These are quite equal to our very good English binding, and are exceedingly cheap. Mr. Möller further offers, if an order be given him to a large extent, say £200, to make similar books at a discount for cash of 20 to 25 per cent. from these prices. Until the duty is repealed, however, 10 per cent. must be added to the price.

It seems that if some plan could be invented for securing the backs of books, without the stiff half-cylinders of iron and leather at present used, a great convenience in writing would be obtained, from the facility in opening, and the much flatter surface they would present to write upon. Knight and Co. exhibited some account books which lie very flat when open; this is caused, apparently, by some degree of pliancy being imparted to the back.

There seems a sort of usage in favour of *vellum* for customers' pass books, otherwise the pliant backs of the books issued by the Savings Banks might be advantageously copied, if they could be written on at all.

An improvement is shown by Mr. Wadderspoon, which, although not new, is little practised, except by some of the first makers, viz., the insertion of linen bands along the sections of books, by which they are prevented from breaking away and the leaves from coming out; this material is useful, also, for protecting indices from wear.

The plan adopted in many banks, of transferring their current accounts into new ledgers every year, has many advantages, especially in the facility which a "dictionary" arrangement affords for posting; but one serious evil is the loss of *continuity in an account,* it being necessary, in case of reference, to move about a number of large and cumbrous volumes.

The thought occurred whether the vulcanised India rubber, recently introduced and exhibited by Mr. McIntosh, might not be rendered available in the binding of a book, by inserting *removeable smooth India rubber bands, in corresponding holes punched through the paper, close to the back,* in such a manner that, at stated intervals, the full sheets might be abstracted and replaced by new. This, or some other plan, might, perhaps, be successfully adopted, and, if so, it seems to be all that is required to obtain continuity of account for any length of time.

An objection, as to genuineness, on the production of a book in cases of legal evidence, might perhaps give way before general usage, should the practice be adopted; and it might be modified by care, as to uniformity of writing and of ink, in carrying forward from page to page.

Account Book Ruling.

The printing of headings and ruling of books, although attended with considerable expense, possesses such advantages, in ensuring accuracy and neatness, that it is well repaid. Our continental and transatlantic brethren, from the system by which they keep their accounts requiring more money columns than we do, have taken greater pains with the ruling of their books.

Considering the great importance of this department of bank books manufacture, it may be well to refer to two or three of the most striking specimens.

Messrs. Gaymard and Gerault exhibited some ponderous books, most beautifully ruled.

Messrs. Carl Kuhn and Sons also exhibited eight books, which for their beautiful ruling and printing are very deserving of notice. In binding, they are, however, very inferior to those from Hamburgh, wanting finish, and their price, ranging from £1 1s. 6d. to £3 10s. 6d., is high, owing to their elaborate ruling and dotted lines.

The specimen book of paper ruling by Messrs. J. and W. McAdams is also very excellent, and is distinguished by the clearness of the execution.

In concluding this part of the subject, attention should be called to the new ruling machine of M. Bauchet Verlinde, of Lisle. This machine is provided with an arrangement by which several sets of lines may be printed at the same time, of different colours, without the danger of their running into each other. It is a modification of what in this country would be called a "drawing ruling machine," as opposed to the "cylinder machine" in common use. By this plan, the usual hindrance, while one colour is drying, is saved, and all the lines running in one direction are at once completed.

Some sheets from an account book were exhibited, measuring 3 ft. 6 in. × 2 ft. 3 in., and containing 328 perpendicular lines of various lengths, in six different shades and colours, and all done at one drawing.

This is not the only superiority in this ruling, the lines are beautifully fine and perfectly unbroken, free from all irregularities. The difference in this respect, is most marked, and is one at which this man has aimed for years to attain, by a superior construction of his "pens,"—they are of a different form to those in use here; instead of the long point in our ruling pens, admitting of continual rubbing and filing, these pens seem as if not intended to be "mended" at all, but are made more in the shape of our steel pens, and with finer ink channels than the ordinary ruling pens, the consequence is, that the lines are such as are not seen in this country, and two or three Paper Rulers have expressed great surprise at the execution. Although such fine lines are not absolutely required for our wider money columns, yet as fine lines look better than coarse they are to be desired; and as writing looks better without any horizontal lines at all, the nearer the approach to a *mathematical* line the better, and these lines of M. Bauchet Verlinde's are almost entitled to that appellation.

Weighing Machines.

The sovereign being not merely the representative of value, but value itself, it is highly important that its intrinsic character should be preserved, and, as a check to the fraudulent practices of "sweating" and "drilling," it is very proper that gold coin should be at all times liable to the ordeal of weighing. The excitement resulting from the proclamation of June, 1842, has long since subsided. For the most part the public look upon a sovereign as worth 20s., whether it be a "horseman" or a "Victoria," and few persons think of weighing gold, except when it is to be sent to the Bank of England, when the new gold is all gathered together, and if enough cannot be found, a few of the best of the old ones are weighed and put with them, the old being considered good enough for practical purposes. It is evident that this cannot go on long, and as some of the gold of the present reign is getting depreciated below the standard, the weighing must shortly be again resorted to. In these days of competition it is found that the seller, rather than displease his customer, is willing, generally, to risk the loss on gold, so that the trouble of weighing and cutting, falls entirely upon the banks. It is desirable, therefore, that this labour should be performed as readily as may be, and that when it is commanded by authority to cut all light gold, all gold discovered to be light should be cut. The practice of bankers weighing sovereigns, and when found to be light giving them back to their customers, is evidently injurious, as it involves the labour of weighing the same money over and over again, until some poor unfortunate brings these sovereigns to take up an acceptance, or on some other urgent necessity, and not having any sovereigns to change for the light ones, is compelled to submit to the deduction. Would it not be for our mutual advantage if all the banks were to unite together to cut all light gold on presentation, in which a royal proclamation certainly warrants them, although it may not warrant their neglect to do so. It is very desirable, before any general weighing becomes necessary, that some improvements should be made upon the scales now in common use, the very best of which are liable to great objections arising from currents of air acting unequally upon the scales, constant irregularity caused by placing and displacing the sovereign to be weighed, by which the equipoise is every moment destroyed, adhesion of the scale-pans to the counter, difference in the judgment of the weighers, failing of the eyesight, difference in the weights of no small amount, considering the great degree of accuracy required.

These and many other objections are overcome, by the valuable and ingenious "automaton balance," or gold weighing machine, invented by William Cotton, Esq., which seems to have resulted from the discovery that the weighing by ordinary scales is extremely unsatisfactory, one clerk rejecting as light that which another accepted as current weight, as is proved by the fact, that on re-weighing three millions of sovereigns in the Bank of England at one time, which had been taken from the public as current weight, one-tenth of them were found to be deficient, and were melted.

This machine, however, beautiful as it is, and suitable as an adjunct to the counter of every considerable bank, is too expensive for any except those who have a very large amount of gold to weigh. They are capable of weighing ten thousand coins per day; the price of these machines was £210, it is now £200, and £10 goes to the Clerk's Widow's Fund, the remainder to Mr. David Napier, who has Mr. Cotton's consent to make these machines. Mr. Cotton very properly took out a patent to prevent others from bringing the machine into disrepute.

There are two or three improvements in the one shown at the Exhibition, and the price of such a one to weigh one size of coin, Mr. Napier says is two hundred guineas; and if required with parts and adjustments, to suit both sovereigns and half sovereigns, the additional expense would be about £30; and he thinks that, to retain accuracy in the weighing, one could not be made cheaper, or of a commoner material. This must be matter of considerable regret, for their value is established by the fact, that the second machine made has been in constant use for six or seven years, and Mr. Cotton believes it has never made a mistake of 1-50th part of a grain. They are in use not only at the Bank of England, in London, but at its branch in Liverpool, and at the Bank of Ireland in Dublin; and as the cashier affirms, they have weighed eighty millions of pieces without a single case of error having been discovered.

There is a slight inaccuracy in the account given of this machine in Hunt's "Handbook," vol. i., p. 377. It should be thus:—"If the piece is heavy, it remains stationary; if light, the beam is raised by the balance, the first hammer goes under it, and the other hammer then strikes it into the light box."

. The best description is contained in the "Illustrated London News" for 22nd March, 1845. And it may be well to give here Mr. Cotton's opinion as to the care required in using it :—"So delicate a machine in course requires attention; dust or dirt on the beam will of course interfere with accurate weighing; unless some person will look carefully to the machine, it will soon be out of order. At the bank, the clerk has two sovereigns, one 1-50th of a grain lighter, and the other 1-50th of a grain heavier, which he passes through the machine, to test its accuracy every morning. In London, the machines are driven by steam, and in Dublin by a weight. If they are driven by hand, it should be a steady one."

. A remark may here be made in reference to many of the articles which have been mentioned, viz., that they must be attended to—that care must be bestowed upon them, or they had better not be introduced. How frequently is it the case that scientific mechanism is rendered wholly useless by ignorance or mismanagement; indeed, the most useful and the best adapted instruments usually require the most constant care.

. It is suggested that some one of intelligence in an establishment should be empowered to do the needful in small matters of renewal and repair, and that every porter in a bank or office should not only possess an *oil bottle* but *know how to use it.*

. The cost of Mr. Cotton's machine is certainly very high, and allowing for deterioration, repairs, and interest of money, would not cost less probably than £30 a year; many, therefore, must be content with modifications of the old scales, and as improvements are suggested by those exhibited, by which some of the objections to the common scales may be lessened or removed, it will be well to refer to them. The complete enclosing of the scales in a glass case, as is done with testing scales and assay balances, would, for the purposes of weighing gold, be quite impossible; but the evils arising from currents of air may be materially lessened by protecting the scales on five sides out of the six; this glazed frame might stand on the further side of the counter, and, running on slides or wheels, be drawn readily towards the cashier when required, and at night a little blind might be drawn down, to protect the scales from dust.

. In the use of scales, the silk cords are very much in the way, and frequently by striking them, a rotatory motion is given to the scale, exceedingly incompatible with perfect accuracy. Perhaps the "balance" action might be preferable, made to work flush, or level with the counter, thus dispensing with the scale strings. The balances exhibited by Beranger and Co., the patent-right of which for England has been purchased by Messrs. Pooley and Son, on account of its delicacy, seems well adapted to this purpose. The patentees are making one, with some little modification, for Messrs. Cunliffe, Brookes, and Co., of Manchester. The working parts being concealed beneath the counter, would be protected from injury, and a little lid put over the scale pans at night would effectually keep them from dust, &c. Perhaps this principle, on account of the increased number of "bearings," is not so well adapted for very exact weighing, but considering that bank scales are scarcely ever in a condition to weigh very accurately, these might really be better suited for practical purposes.

An improvement which might be adapted to these balances, and which might rectify the errors arising from an occasional failure of sight, is suggested by the position of the index rod in the assay balances and testing scales which are exhibited. In all of these, the rod is brought from the scale beam downwards, and the index plate is fixed to the bottom of the pillar. In gold-weighing scales, the pointing rod is always put above the beam, so that at each action the eye has to quit the scale, and be raised to the top; this is very fatiguing, and no advantage at all seems to be gained by it; but the evil attending it is, that the eye is wearied, and the exact position of the scale, is frequently missed in placing the coin upon it, thus causing vibration, and the equipoise is lost.

The importance of an index and pointing rod at the bottom of the pillar was suggested to Messrs. Chambers and Co., who intend shortly to bring out something of the kind for money scales.

Messrs. Nicholl and Co exhib ted a very superior balance constructed in this way, with indicating dial half way down, though not intended for weighing coin.

M. L. Reimann, of Berlin, exhibits some delicate scales, and one of them is provided with a screw, by which the scales are raised, thus ensuring greater steadiness than can be obtained by the application of the finger to the lever; but much gentleness of touch is acquired by practice, and those with a lever and a little wheel raising a sliding bar are perhaps the best scales in use.

Lelune and Co., of Berlin, also manufacture some beautifully delicate scales, with an index at the bottom.

Decimal Balances.

Much time and trouble would be saved by the use of decimal balances, of which Mr. Broemel is an excellent maker. He says that 4620 patent weighing machines have been made at his establishment in the course of eighteen years; one beautifully executed little decimal balance, which was stated to weigh 100 lbs (of course with a 10 lbs weight) is very deserving of attention. This would be just the thing for bankers' counters. Price, £5 5s.; and with it, the weighing of gold in bulk would be greatly expedited. For instance, in weighing 1,000 sovereigns, it is usual to count 100 first; then follow three shiftings from scale to scale, and four weighings; or if you begin with the 100 sovereigns weight, then five weighings and four shiftings are required; whereas with the decimal balance, the operation is performed by once raising the scale, with the 100 sovereign weight on the one side, and the gold on the other. This would do very well for full weight gold.

A letter balance of Messrs. De Grave and Co. deserves notice in this place; this method for letter weighing seems far preferable to the scales in ordinary use, the letters being of various shapes and sizes, are not conveniently placed within the cords of a scale—less still so upon the balance plates with a single suspending rod; the *balance* surface on the other hand admits of any thing being placed upon it of whatever shape, and being fixed, the parcel is not in danger of falling off.

In concluding this division of the subject, it may be desirable to call attention to the propriety of provincial banks employing provincial tradesmen in the manufacture of account books and the provision of stationery materials. In all districts of the country the trades and professions are supported in a measure by each other, and it is manifest that the prosperity of its own locality is of more importance than that of the metropolis.

The practice of sending to London for stationery and account books, which are the chief if not the only articles consumed by bankers, as such, of course arises from an expectation of being better served; and with the constant applications of London houses, of high respectability, for orders, presented through travellers, with the most polite assiduity, it is perhaps difficult to do otherwise. The consequence, if immediately beneficial, is ultimately injurious; it causes and perpetuates the evil which is named as a reason why London firms should be employed, viz.—that a good book cannot be made in the provinces; the reason of this is manifest, that a good binder cannot get a living elsewhere than in London. The country towns and the cities also become, as it were, importers, instead of exporters; the labour is employed elsewhere, attended by the profit, and is of course followed in a measure, by the capital, drawing all after it. The intelligent workman who has served his time, and

saved his money, and who would, in ordinary cases, attempt to start in business for himself, with recent plans and new machinery is discouraged, and
goes to one of the great London houses; the effect of this is that *tradesmen*
suffer; they are obliged to send to London through their stationers for books.
London has a monopoly, and prices are raised. Thus the stationery business
of a county is affected by the arrangements of a single establishment. It
will be remembered that a bank with ten or twelve branches, spending
between £200 and £300 a year in stationery, would, with a few other principal
establishments, be sufficient to support a first-rate man, and enable him to
procure adequate machinery, and this is one principal difficulty, the expense of
machinery, which country tradesmen cannot meet, unless they have all, or
nearly all the work. As a simple illustration, one may generally tell a country
made book, by observing that the edges are cut by the old horizontal *plough
knife*, instead of Wilson's excellent perpendicular cutting machine, Class 6,
No. 112. This, and many other valuable facilities in account book manufacture, including folioing machines, and powerful presses, are only to be
obtained by tradesmen who have a considerable amount of business. The
obvious course would appear to be to begin with those books which are least
important, and order these in the country, also all articles of general stationery.
It will be found that the paper may be just as good, and the materials generally;
only there will be less "finish," about the books, and the "lettering" may be
a little irregular—this however will improve. Then in reference to important
books, those requiring to sustain *long wear*, and some which it is desired should
be especially good, they may be ordered at the provincial establishment, just
giving a hint that they must be *Thomas's* make, this will be understood; the
provincial tradesmen will share with them in the advantage, and there will be
no difficulty in recognizing the book when it arrives, even without their label.

Letter Copying Machines.

At a period when all copying of letters had to be performed by hand, and
by the process of transcription, some solicitors and commercial firms found it
answer their purpose as a rule, not to copy their letters. Now, however, when
the facilities of presses are so great, the least troublesome, and the safest way,
is to copy every thing, this saves much time in consideration, and with proper
arrangement of several books for distinct classes, letter copies may be kept in
a very fairly accessible form. Since the adoption of the penny postage, the
letter department in banks has of course been much increased, and the copying
of letters assumes much of importance as a daily labour, and should be in
every way facilitated, otherwise some of the advantage will be sacrificed in the
delay.

The merits of the various machines exhibited may be noticed in three
divisions.

1st.—Those the construction of which is calculated especially to save time.

2nd.—Some combinations by which more power is obtained.

3rdly.—The form most calculated for endurance.

Some of the presses unite all these, but are perhaps generally more distinguished by one particular.

1st. The press exhibited by Mr. Perry is almost the only one upon an entirely
new principle, and combines very considerable power with great rapidity of
execution. It is a combination apparently of the mechanical powers of the
lever and the inclined plane. The pressure is obtained by means of a lever
handle of moderate length, by suddenly bringing into an upright position two
leaning bars—thus their length is, in the position which they occupy, practically increased, and this increased length, is at the expense in the thickness
of the copy book, which is proportionably diminished. This possesses an advantage in the immediate application of the power, and is better than the
ordinary lever presses, as the horizontal action of the handle, which is in the

same manner as ordinary printing presses, is less fatiguing to the hand and arm.

In the ordinary lever presses, the surface between the handle and the perpendicular pressure bar, from extreme friction becomes much cut and worn, by which the friction is much increased and the handle becomes liable to *fly up*. It is to be feared that this press is carried back to Montreal, but it might answer the purpose to obtain some from thence, even against a 10 per cent. duty, if upon trial they should be found to answer. The price Mr. Perry asked for it was £5.

2nd.—The principle of the press exhibited by the Coalbrook Dale Iron Company, with double screw, is capable of exerting very considerable power, and may be useful perhaps in copying ten or twenty letters at a time, and if thought desirable to keep two presses, in case of one getting out of order, this might be tried, but the mechanical principle, "that what is gained in power is lost in time," is here most fully exemplified. The construction is not nearly so massive as seems to be required by such a principle, the price however is only 55s.

3rd.—The construction of some of the presses which attract attention under a third class, are those which seem specially adapted to wear long, in connection with other valuable qualities. "The newly Improved Copying Press," made by Barrett and Co., is very well and strongly made. Its principal advantage consists in the diminished friction at the point of pressure, where instead of the two rubbing surfaces, a wheel is introduced, this greatly facilitating the application of the power, and rendering the press more lasting; as in all machinery where there is undue pressure on very limited surfaces, there must be corresponding wear. It is found practically important that presses whether screw or lever should be *fixed*, their weight being insufficient when used rapidly, and of course this press, from the lever pivot being on one side, and the lever being longer, will be more likely to tilt up than those with the lever handle working in the middle; the prices of these machines without the "et ceteras" are moderate—those $9 \times 13\frac{1}{2}$ inches £3, those 11×18 inches £3 10s.

There is one important feature in Pierce's Imperial Copying Press, viz.,— that the pressure is not exerted upon the machinery of the press during the whole time the letter is in process of transfer, but merely for the moment of applying and removing the power, in order to place and displace the book; this is calculated to make it last much longer; the construction is extremely simple, the two plates are connected together by four stout rings of vulcanized India rubber, two at each end; the lever is employed in the same manner as in Messrs. Barrett & Co.'s press, to separate the plates; the copy book is then inserted, and when sufficiently pressed the power is again applied and the book removed. The price of this press is such as to recommend a trial of it. For press 11×9 inches 30s.

Screw Copying Presses.

One prominent defect marks the majority of the screw copying presses, and after being in use for a year or two they remind one strongly of an exploded steam boiler, or a young blind horse. They are too powerful for themselves, and their construction does not recognize the mighty principle which they contain, and thus in the attempt to make them ornamental, their efficiency is sacrificed; and although perfectly firm when new, the nuts, however tightly screwed, will move, and the pillars shake and totter, and this notwithstanding frequent attention and repair. As a striking exception to this, however, the presses of Messrs. Ransome and May are peculiarly distinguished for solidity and strength. The lower plate and the arch in which the screw works, are cast in one piece; this imparts a degree of firmness which cannot be obtained by screws and nuts. But even these do not seem to come up to the mark of what is required for actual every-day, and all-day-long service. The screw should be larger, to prevent its wearing out in twelve months, and the whole thing still more massive. If the handle ends were heavily weighted also, the screw, if well oiled, with one pull would work itself, and save much manual labour. Of course, these remarks are not intended to apply to ordinary presses, or for branch banks, where only a small number of letters may be copied daily; but where the number is considerable, some improvement is necessary. Perhaps it may be considered that the screw, under favourable circumstances, is the most useful application for letter copying, remembering the advantages it affords of regulation in pressure, according to the time available for procuring a copy; and of screw presses, those lastly mentioned are certainly the best. Prices :—folio, £3 5s.; large post, £2 7s. 6d.; and three other sizes.

The presses by Messrs. Shenstone and Mills are deserving of distinction, on account of their style of execution, especially the large folio press, price £4 4s.; and although elegance may not be of much importance in such a utilitarian commodity as a copying press, it is but right to point out the very beautiful workmanship of those by M. L. Poirier.

Paper Damper.

There seems to be at present some difficulty in the process of damping the copying paper. The plan of sponging a plate of canvass-covered copper, and then pressing it first in the book to moisten the sheet, is attended with twofold trouble, and to avoid this, the plan has been adopted of sponging the paper itself with a flat-sided sponge, and then drying with a fold of linen. An ingenious instrument for avoiding this loss of time is to be obtained of Mr. M'Corquodale, whose patent damper consists of a hollow iron cylinder, about a foot long and an inch and a half in diameter, in which is a piece of cloth. In the cylinder there is an aperture the whole length the thickness of the cloth, which is by a screw at the end capable of being drawn out or in; it generally protrudes half an inch or so. Near the centre of the cylinder is a small air hole, protected by a key, which, on being raised, admits the air into the cylinder, and this imparts wet to the cloth, which is then drawn over the copying paper; on the key being closed, the wetting of the cloth immediately ceases, being air-tight. The damper costs 10s. 6d. Perhaps this is not the best plan which could be devised for the purpose, and it seems important that press manufacturers should turn their attention to the subject.

Patent Autographic Press.

Messrs. Waterlow and Sons have recently patented an invention under this name, which professes to afford great facilities to bankers for multiplying copies of letters, circulars, &c. It appears to differ from the old lithographic system, in the transfer being made to a highly polished metallic plate, instead of the lithographic stone hitherto used, and the press is in the form used for copying presses some years ago. The great practical difficulties with the old lithographic press were, first, in writing upon the transfer paper, requiring a special pen and ink and special care, it being somewhat of a greasy nature; then, secondly, considerable practice was found to be necessary in successfully transferring the subject to the stone; these objections to the old, may perhaps be quite removed in the new system adopted. It is, however, rather remarkable that in the testimonials published, they are entirely " opinions " of the "press," and newspapers are not, upon such matters generally, a very practical authority. As this press has now been before the public for fifteen months, it may be expected that some testimonials of its worth from those who have used it, will shortly appear.

The type-paging of letter copying books by steam power by Messrs. Waterlow and Sons, will afford a great facility in indexing and in reference, the date being longer to write in the index than the folio; and for reference, the date of the letter not being close at the corner of the sheet, and many letters being copied in one day, the difficulty of finding any one is very considerable.

Manifold Writer.

On ordinary occasions, it is desirable that letters should not be copied by the same party who writes them, the process being purely mechanical; but for the Manager's private correspondence, "Wedgwood's Improved Manifold Writer" would be very useful; this requires little arrangement. The registered clip desk would add to the facility. It may not be generally known, that with care and practice, as many as six copies may be written at once by this means. The smooth glass style seems well adapted to this purpose.

Two or three other instruments used in carrying on the business of a bank, will now be pointed out, including

Stamping and Printing Presses, Seals, and Letter Clips.

The "crossing" of checks may be considered as an important operation, both as regards the prevention of fraudulent misappropriation and the facility it affords in the tracing of money transactions, which is frequently of use. The block upon which the printer's ink is sometimes insufficiently "distributed" being formed of a composition of treacle and glue, generally used by the printers for their rollers, is liable to become hard, and unless kept scrupulously clean and frequently renewed, the ink adheres to it, and the impression from the stamp becomes thick and unsightly. A very general defect in the impression thus produced by the hand is, that there is too much ink and too little pressure; they become rubbed as soon as pressed, and the checks are subsequently very unpleasant to handle. As a general rule, it may be said, that for stamping checks the ordinary printer's ink is too thick for hand pressure. The printing apparatus registered by Harrild and Sons is well adapted to supersede the old plan of stamping by hand. It is self-inking, and the raising of the handle causes a composition roller to pass over the type; and then, when the handle is brought down, a leverage is obtained, causing a very effectual impression. The price of this, if complete, is £2 6s.

The press made by Schlesinger and Co. is superior to the last-named in some respects. The arc of the circle described by the hand in working it, is not much more than half the former, viz., about 45 degrees, by which much greater rapidity of action may be gained; the composition of the roller is peculiarly adapted to its purpose. The *printer's composition* is only suitable where a constant and active movement is going on, when, in fact, it is in full work, and where it can be well washed and renewed repeatedly. In this press, the inking roller is of a different construction; it is a hollow cylinder, bound with cloth on the outside; there are some fine communications between the inside and the outside, the roller being filled with fluid but substantial ink, it gradually works through to the outside, as the demand is made upon it, and the makers say that, for ordinary use, the roller will contain a six month's supply. By raising the handle, the roller is brought in contact with the die; by pushing down the handle, the impression is then transferred to the paper. The price of this machine is £3 3s., complete, with die not cut; and any sort of stamp can be made to order for it, cut in brass, at 3½d. per letter.

It has often been remarked how well all the foreign bills are stamped and marked; and the manner in which they make their stamps is very peculiar, the letters are not cut out of the solid metal, as ours are, but they are built up with fine thin plate brass or copper, with beautiful neatness and regularity; the outline of these letters is most acute, and those brought over from Vienna by Carl Dinkler are especially beautiful.

Consecutive Numerical Stamping Machines.

Now that the paging of books is so small an addition to their ordinary cost, the best way is to have them paged by the maker; the same may be said of notes, lodgment receipts, checks, letters of credit, &c. There are, however, some other purposes for which these machines may be found useful, amongst which the stamping of bills of exchange may be noticed; in some banks, where a great deal of discounting is done, it might be as well to have such a press as Harrild's or Schlesinger's, both of which are very compact and convenient, but the expense would be a heavy consideration, viz., £25. Harrild's combines the numbering machine and the above-mentioned stamping machine in one. For those banks which issue Bank of England notes with their own stamp upon them, the " Stamp Counter " apparatus, invented by M. C. de la Baume, may be useful; by this means, the number of notes which had been stamped would be checked; also, in despatching letters or circulars, the "date on which posted" might be stamped in the corner, just before placing in the post bag, and the number indicated by the dial, then compared with the number in the dispatch book, should correspond; this would be a check upon all the letters written being actually dispatched. The expense is very inconsiderable, without the die £2, and of course one or many dies might be adapted to it.

The folioing machine sent over from Boston by W. and J. M'Adams is deserving of notice, on account of its extreme singularity. It consists of a long chain of type numbers, each group apparently cast in one piece. This chain, which is perhaps twelve feet long, is made to revolve round rollers, a link, or number plate at a time, by a motion of the foot, and the whole chain having been inked, at each movement an impression is produced upon the book or paper placed upon the platform. Unless the type can be very cheaply made, it would seem that this must be an expensive construction, although some of the apparatus is dispensed with.

Embossing Presses.

It would be well if some way could be discovered by which the exceedingly neat and clean impression of these machines could be rendered permanent and irremoveable. At present, although exceedingly adapted in some respects for stamping cheques, the possibility of obliterating the impression by pressing and rubbing is a decided objection, as a "crossing" once made should not be liable to be erased in any way.

As at present used, they may perhaps be applicable for some purposes, and the suggestion having been made to some of the manufacturers that permanence in the impression was very desirable, it is quite possible that some improvement may be the result. Two or three makers may be mentioned, whose presses manifest considerable care and judgment in their construction. Mr. Muir's India-rubber spring press is very much adapted for rapid use, the weighted end of the handle merely requiring a single pull from the hand, this causes the pressure, and the spring immediately returns the screw to its former position. The whole of Mr. Muir's assortment seem constructed in a very scientific manner, which may also be said in reference to the embossing presses exhibited by Mr. Jarrett, Messrs. T. Wells, Ingram and Co., and Mr. Collett.

Seals.

The watch-ribbon and pendant seals are now but rarely seen, and the enormous wax impressions upon letters have almost disappeared. These have been replaced by smaller seals, and very recently in a majority of cases, by adhesive gum.

The latter, however, is not suitable perhaps for many occasions occurring in banking business, where there is either security for value to be ensured, or secrecy in reference to matters of a confidential character, necessary; and there are means of opening adhesive envelopes, which make them sometimes unfitting enclosures. In many cases, however, the adhesive envelopes, or letter paper, which might be made as readily adhesive, may be used, and now that letters are carried for one penny, it is desirable that every facility should be afforded for their convenient transmission; and a thick seal, whether small or large, is a most inconvenient thing in a letter carrier's budget. Seals, too, if well mixed, take much time to make. Even whilst regarding it as a duty to diminish by all means the unnecessary thickness of letters, it is difficult to reconcile oneself all at once to the abandonment of the highly ornamental wax impression; and the temptation is irresistible to call attention to the rare specimens of skill exhibited in Mr. Martin's "Tornography." This is certainly a most beautiful manner of cutting seals, and the sharpness and depth of the impression are highly ornamental; but, as has been already said, the gratification at this invention must, however, be qualified by the fact, that the post office (to whom the public are greatly indebted) does not like thick seals; and as a matter of security, a thin, well pressed seal is always a greater protection than a thick one, for there are curious devices for removing seals, which are very difficult of execution where there is little depth of wax.

Letter Clips.

Much convenience may be found in the use of letter clips, not so much for confining letters as for holding checks during the entry of them in the day book, &c. The process may be considerably assisted thereby, as they may be turned over readdy with the left hand, without fear of damaging the position of the pile; the most convenient plan will be to confine them at the upper left hand corner with one of Whitehouse and Sons' Registered Square Clips, with spring and wheel—price 1s.

For confining letters, "Circulars to Bankers," or pamphlets, the clip book exibited by De la Rue and Co. would be very suitable; or, better still, the screw leaf-holder by Charles Knight.

VI.—IMPROVEMENTS IN PRINTING AND ENGRAVING.

ENGRAVING AND PRINTING CHEQUES.

The engraving and printing at present required for a bank note and for a check differ somewhat materially. As has been already remarked, the danger of fraud with regard to checks consists not so much in the imitation of the engraving or the paper, as in the alteration of the "amount;" this offence, although amounting to the crime of "forgery," being more lightly punished than the imitation of a signature. For the purpose of preventing this, therefore, the "vegetable colour" is well adapted for checks, whether embodied in the pulp from which the paper is made, producing the "tinted" check paper, or composing the ink with which the check is printed; therefore, for all purposes of safety, a plain check plate is practically as good as an elaborate one.

Some remarks upon the tinted paper will be found under heading "Paper for Checks."

The plan adopted by some banks of having their checks engraved, and then the name of the branch, &c., inserted in letter-press, is rather an expensive mode, requiring a double process. Additional prominence is frequently desirable in some parts of an engraved or printed document, and this is not always sufficiently obtained by a greater size of letter, but is more effectually produced by the use of a different colour; this has usually necessitated two processes, but in the samples of printing exhibited by Mr. Mackenzie, two distinct colours are impressed by one "form" at one time. This is done by using materials of two different heights.

An invention, corresponding somewhat in effect with the last named, though very superior in principle, has been made by Mr. Fisher, the inventor of the postage stamp colour, &c. The specimens exhibited are apparently printed in two colours, a neutral tint for the back ground, and black for the letters; and

D

yet these two are produced at one printing, from one plate, at one press, and with one inking. This is certainly very singular. It is caused by a peculiar property in the colour used, which is extremely sensible to alteration by the quantity placed upon the paper. The engraving is on steel, and in the process a *back ground* is worked upon the plate extremely fine, the *letters* being engraved much more heavily; when finished, to all appearance it presents one smooth and even surface, and is inked at one operation. This invention possesses a three-fold advantage. 1st. The greater distinctness in the impression, having two colours without additional labour. 2nd. The vegetable colour being highly sensitive, prevents its being tampered with by acids, &c. 3rd. The secret composition of the colour would prevent the imitation of documents printed in it. The first and second may be considered as bearing upon the printing of *checks*, the third may be valuable in reference to *notes*, the object being to preclude their *imitation*, rather than prevent their being altered.

ENGRAVING AND PRINTING NOTES.

The usual custom of engraving notes in a manner extremely elaborate and minute, has some advantages, and the end in view, viz., to prevent *correct imitation*, is certainly obtained; for there are very few engravers capable of such execution, and those few are too fully employed by Messrs. Perkins and Co., and some of the leading tradesmen, to find any time for mischief; the ultimate object is, however, to prevent loss, and in this particular the minute engraving might not be equally successful. This minute engraving partakes rather of the character of the secret marks of the Bank of England notes, and to themselves alone these marks may be very useful, and are so, for it is very seldom that the loss of forgeries falls upon them.

It seems desirable that in bankers' notes, the minutely engraved groundwork sometimes seen, and consisting of the repetition, hundreds of times over, of the name of the bank, should not be entirely relied upon. Not one in one thousand of the public, perhaps, is aware of its existence ; much of this work is hardly legible by the naked eye, and a person unaccustomed to scrutinize it at the counter, would not observe its absence. For, the purpose, therefore, of the public's participation in the advantage the banker secures to himself, it might be well to accompany this minute work with a fully engraved note on both sides, in such a manner that persons of ordinary intelligence may judge whether they have a good note in their hands or a worthless strip; thus the value of our paper currency will be enhanced. Mr. Fisher's ink colour referred to above, will be very suitable for printing the notes, by this means additional prominence may be given to the amounts, &c., although its susceptibility as a vegetable tint would be most available for checks. Mr. Fisher has not, at present, introduced his colour for sale, being desirous to dispose of his invention.

The patent process exhibited by Messrs. Royston and Brown, will be very useful for minute engraving; the charge for 20,000 notes of the usual size, in strong bank note paper, including extras for water mark, numbering plate, &c., would be about £3 15s. per thousand, a larger number would be proportionably less.

Cutting Notes.

In connection with the engraving of Notes, a reference may be made to the very considerable expense attending their circulation ; of course the advantages derived from a paper currency are mutually shared by bankers and by the public, and the interests of both are closely identified. It is to be regretted, therefore, that the apparent interest of one should in any way be sought at the expense of the other, as in the case of " *cutting notes*," the expense of prematurely renewing these, the engraving, paper, &c., being alone considered, is something; but when the trouble of cancelling, writing off, entering, and re-entering, examining of numbers, signing, counter-signing, dating and checking, is all considered, the labour is very great indeed. It is difficult to account for the great number of cut notes, now that commercial travellers so uniformly take drafts on London, unless it be that some bank, with a very large number of branches may have thought it right to adopt this plan for safety, bankers being, of course, obliged to send notes by post. There seems to be an idea upon the public mind, that if half a note comes safely, that is all they require. It may be well for them to understand that bankers do not pay single halves of notes *upon demand*, and that drafts on London, or even post office orders are a far preferable mode of remittance.

Notes in Duplicate.

In considering the subject of cutting notes, and the amount of trouble caused thereby to banks of issue, in some localities, the question has arisen, what would be the amount of inconvenience attendant upon the issue of paper in duplicate, and what would be its advantage? The additional trouble of printing would be very small, as they need be no larger than at present, and in fact, printed at the same time, and cut afterwards, only requiring a completed margin; indeed, the present plates might merely be divided; the number is at present in duplicate, the date is frequently so, the signature and counter-signature are on different halves; in fact they seem as if intended to be divided. An objection is, however, immediately raised, that the labour of " telling " will be double too. This, however, may perhaps be overruled and be found to possess an advantage. No cashier thinks of issuing notes without telling them twice, and that at opposite corners, thinking it possible that an adhering note may be *twice* passed in the *same position*, and so it may, and occasionally *thrice*; if, however, the cashier has to pay £60 out of a £100 parcel of notes, he would usually be content with counting the £60 once, leaving them to be checked by the customer, and having counted eight remaining notes confidently replaces them in the drawer.

Supposing the pile of notes to be cut in two, and placed in two separate divisions in the cash drawer, a check being presented, and the notes counted once from each pile, a comparison of the number at the top of the piles would be an equal security with the transaction just described, with this additional advantage, that a note may be more easily passed twice in the *same*, than twice in *separate* parcels; the smallness of the notes would, of course, be inconvenient at first, but the plan which might be adopted of throwing them up separately with the thumb from the counter into the hand, by which the light is seen through them, in a measure, and by which a more extended contact with the paper is obtained, would be soon more speedily performed than the ordinary movement of holding the parcel with the left hand, and raising the corners with the right. The principal objection to this mode would be in the necessity it would involve for the comparison of notes received at the counter ; this might not be very great, it being insisted on that the halves be presented in corresponding order, they might be set aside until a leisure moment, when, with each hand the separate halves might be turned over, and would then be fit for re-issue.

VII.—NEW INVENTIONS IN THE CONSTRUCTION OF LOCKS AND SAFES.

The recent " lock controversy " has much increased the public interest in this subject, and some may perhaps expect that much will be said in reference thereto; but it is conceived that the abundant information already elicited by that interest from abler writers on this subject, renders any further disquisition thereupon unnecessary. It will be right, however, to allude to some of the locks which seem to occupy a prominent position, either from their superiority of construction, or novelty of design.

It is to be hoped that the mechanical spirit which has recently been awakened may be productive of still greater efforts in the improvement of this important piece of mechanism for protecting property. Lock picking will henceforth be more fully contemplated and provided against, in connection with lock making. For this we may thank our brethren in America. And, perhaps, it is fortunate that the picking of the Bramah lock has been accomplished at so opportune a moment by Mr. Hobbs, through whose courtesy some light has been thrown upon this mystery, in the lectures at the Banking Institute and before the Society of Arts.

The circumstances under which the lock was opened are well known. It was in the hands of the operator sixteen days, who made use of a fixed apparatus, screwed to the wood-work in which the lock was enclosed, together with the assistance of a reflector, a trunk of tools, and four or five other instruments made for the purpose, having been allowed six or seven weeks previously to take wax impressions of the key hole. The only legitimate way for Messrs. Bramah to obtain "satisfaction" from Mr. Hobbs, will be to pick *his* lock, U. S. A. No. 298, capable of 1,307,654,358,000 permutations. For a time, Mr. Hobbs' locks (apart from their expense) will stand pre-eminent; considering that the lock picked was fifty years old, should a second challenge upon similar terms, in reference to a modern lock, be declined by Mr. Hobbs, the case will then appear materially altered. Mr. Hobbs will, of course, be willing to offer similar facilities for opening his prize locks, and if Messrs. Bramah cannot accomplish this within sixteen days, it may be probable that they cannot accomplish it at all. There can be no doubt that the principle of Mr. Hobbs' locks is most excellent, and their execution is highly creditable to America, whose locks at present stand A 1. It will be remembered, however, that his best lock is £50, whilst Chubb's best lock, with 30 tumblers, is only £15.

Some good must certainly result from this " controversy," and probably, the credit of the Bramah principle and execution will in no way suffer, but eventually be advanced by those improvements which will be consequent upon the fact of a highly accomplished American artist having succeeded, after an arduous struggle, in opening them, the facilities afforded him being such as no thief could ever possess, even if he had the necessary ability.

The case, however, has been one of no ordinary picking, and we are much indebted to the perseverance and ability of Mr. Hobbs for the stimulus which, in all probability, will be given to lock manufacture in this country.

The Messrs. Bramah will, of course, improve their lock, and increase its present high reputation. Indeed, the position of the vanquished in *such* a struggle is far from discreditable.

Whilst all locks should be distinguished by security, strength, simplicity, and durability, a very large proportion are not intended, and need not to be constructed with a view to resist actual violence. This class is useful in the prevention of petty fraud and prying curiosity, and is applicable to all those cases where the certainty of detection would deter from robbery. For the security of valuable treasure, another character of lock should be employed ; one calculated not only to resist the secret attempts at tampering, but also the desperate application of main force. The discovery by the metropolitan police, in the early part of 1845, of the burglar's instrument called the "jack in the box," may well shake our confidence in the apparent security of an iron strong room door. This instrument is so small in compass, that it might be easily carried about the person, and yet it has the power of lifting three tons weight; this is accomplished by the power of the screw, a turned iron being inserted in the key hole, a purchase is then gained up in the surface of the door, and either the lock or a portion of the door itself is torn away. Another plan occasionally adopted for opening safes is by the insertion into the key hole of a burglar's "brace;" this instrument is then forced round, the lock is by this means entirely destroyed, and then the bolt is readily shot back by some instrument, or sometimes by gunpowder. In some cases the "stub" which holds the bolt when it is thrown has been drilled out (its position being generally known); the bolt is thus at liberty. In other cases, burglars have avoided the door altogether, and obtained an entrance to the strong room by excavation; time, therefore, being granted, it is possible for men to get through almost anything, therefore in addition to locks, and bars, and bolts, and doors, it is essential that some one or more individuals should be day and night on the premises, wherein property to a large amount is kept. A secure room should be arranged, as near the safe as may be, wherein some one should sleep, with closed and fastened doors; fire arms forming part of its furniture. The room also should be so constructed, that the shutting of the door may not impede the ready perception of the slightest noise: a person in such a case cannot be suddenly surprised by any one concealed within the building, but would have time for thought and preparation should he be disturbed. In addition to this arrangement, of course the strongest available defence in stone and iron should be used; but it is evident, for the protection of valuable property, they alone are insufficient. A useful addition to the safe door locks will be found in the sliding rod, occasionally adopted, which may be lowered from the room above through the safe door.

Some delicate machinery has been constructed for the purpose of alarms, which are frequently attached to doors and windows, bank-safes, and even *locks.* It has, however, been found, that the trouble of nightly adjusting all these instruments, together with the attention necessary to insure their action, and the annoyance caused sometimes by *false* alarms, form a serious objection to their use, and, it is thought that if a safe is sufficiently secured, no successful effort could be made to enter it without the person sleeping near it being aroused, especially if aided by a little dog.

It may be well to notice in this place the peculiar suitability of the locks constructed by Mr. Marr, for all safe-doors; these are so constructed that they cannot be affected by the burglar's brace, and no one can open them, even if entrusted with the keys, without being instructed as to the secret of doing so. In Mr. Marr's arrangement, two locks are contained: one is placed behind the other; the hinder one serves to lock the bolt, it having been thrown by the handle. The outer lock, with a second key, then throws a strong hardened steel plate over the key-hole. This is far preferable to the ordinary brass escutcheon locks. The key-holes are placed at right angles with each other, thus making the introduction of picklocks impossible; and the outer key-hole of the door is at right angles with the outer lock. The keys are small, thus preventing the facility afforded by large key-holes for the use of force or cunning. Mr. Marr has very wisely abstained from advertising the principle of his lock, thinking, the less that is publicly known of it the better. In the construction of safe doors, Mr. Marr protects his locks from being drilled, by riveting a number of old files on the back of the door, thus gaining additional strength.

Mr. Hobbs has not had the opportunity afforded him of earning 200 guineas in sixteen days, by exercising his talents upon the locks of Chubb and Son. A lock, however, bearing their name has been picked by him in the presence of Mr. Porter, of the Board of Trade, Mr. Galloway, and other engineers. Messrs. Chubb affirm that this was an old lock of their father's, made under a former patent, without the modern improvements. One particular feature in the Chubb lock is the "detector." This consists of a simple arrangement which is brought into action when either of the "tumblers" is overlifted by a false key or picklock; when this is done, the true key will not unlock it until it has been released. This is done by "reversing the key." This detector system, which is adopted in many locks, may be useful for the discovery of fraud, but the *ready manner* of releasing the detector appears decidedly objectionable, as, on attempting to unlock the door by the proper key, the hindrance would probably be thought to be accidental, or arising from some misplacement, and in a moment, without reflection, the key would very probably be reversed, and the idea never occur that the detector had been thrown at all. This principle of reversing the ordinary key, in order to reinstate the detector, appears to be adopted in all the detector locks, and its extreme readiness appears a great objection.

The plan adopted in the Chubb locks at the Westminster Bridewell, where 1,100 locks are fixed, forming one series, is far superior, when, in case of any surreptitious attempt being made to open a lock, and the detector being thrown, the governor alone has the power, with his key, to replace the lock in its original state.

The detector arrangement, although possessing some advantages, is not without its evils, especially in locks where the detector is liable to be thrown by the tumbler being very *slightly overlifted*, as in this case, the pressure of the detector commencing almost immediately upon the tumbler being raised to its *proper elevation* for allowing the bolt to pass, may indicate to the lockpicker the character of instrument to be used ; and in such locks, when by long use the tumbler springs are considerably weakened, the detector may be sometimes started by a sharp movement of the *proper* key. Where there exists the remotest possibility of this occurring, the detector were far better absent, as the discovery of a detector being thrown is one of grave importance ; this danger will be obviated by placing the detector some distance above the ordinary range of the tumblers, and although not quite so much skill may be required for its manufacture, yet the lock exhibited by Tann and Sons, which requires that the tumbler should be raised considerably beyond its proper height in order to throw the detector, may be, for that reason, the more really useful lock. It seems that detectors generally should neither be too readily thrown, nor too readily adjusted.

Guarded Tumbler Locks.

Messrs. Tann and Sons' locks are also provided with a "flange," or "guard," affixed at right angles to the edge of one or more of the "tumblers," thereby covering and entirely protecting the *edge of the "tumbler," above it* from the action of a "pick," and supposing all the "tumblers" but this one to be successfully raised, the bolt could not be moved.

It is very desirable that keys should be made as small as possible, consistently with the power required, that they may be conveniently carried in the pocket; and these locks are distinguished in this particular, the heavy bolts being first shot by a handle in the door.

Where the workmanship of locks is very fine, small keys can be made to shoot large bolts, as is beautifully exemplified in the lock exhibited by M. Grangoir, 2 inches thick, 10 inches long, and 6 inches deep, having a bolt $1\frac{3}{4}$ inch by $1\frac{1}{4}$ inch ; this bolt is shot out two inches at two revolutions of the key, the key being only 1 inch long and $\frac{1}{8}$ inch thick, the handle ring $\frac{1}{2}$ inch diameter, and the stub of the key is only $\frac{1}{16}$ of an inch long; the whole being not longer than a small watch-key.

The lock exhibited by Messrs. Barron and Son, with eleven tumblers, appears to correspond very materially in principle with those by Messrs. Chubb; the original patent being, however, for a lock with two tumblers moving in a racked bolt, these tumblers placed in different radii, and the key "bitted" accordingly. In both cases, the bolt is released by the raising of tumblers to a certain height. The detector is also regulated by the proper key, when it has been thrown.

The detectors of the locks exhibited by Mr. Gibbons are upon the same principle. Those made by Mr. Woolverson are distinguished by the delicate poising of the detector, by which it is rendered very susceptible.

An apparently superior method of "detection" is afforded in the lock constructed by Mr. Huffer. If a false key is inserted, it is immediately secured by a revolving wheel immediately behind the face of the lock, closing the key-hole entirely; the lock is provided with a second key-hole, having a series of "sliders," which are operated upon by the true key to release the false one.

In this lock, and also in that exhibited by Mr. Foster, there is a curious arrangement for "protection" as well as "detection," for, on attempting to tamper with them, *lancets* are shot from the sides of the key-hole, calculated to inflict considerable injury on the hand that would invade it.

Mr. Cotterill's locks are made very scientifically, being, in fact, fitted to the keys, which are all cut unlike one another; the key-cutting machine, Mr. Cotterill asserts, is constructed on a scale of a million to the inch, two keys only being cut whilst the machine is in one position. If the locks are made upon a scale equally exact, it would appear that the alteration in the size of the keys,. from *variation of the weather* (!) must inconveniently affect their action, and that a key which, at a temperature of 40 degrees, would pass the lock, would, if raised to 70, throw the detector, which is "easily" released in the ordinary way. It is very desirable that keys should be varied in size although the value of the difference does not consist in its minuteness.

Mr. Cotterill has some very high testimonials from practical engineers, and machinists, in reference to the general construction of his locks; and from the indentation in the key being so varied in depth, and inclination, it would be extremely difficult to take an accurate impression.

The locks exhibited by Mr. Taylor, called "Improved Balance Detector Lever Locks," have no detector, although called by the name (and perhaps they are as well without). The tumblers are capable of variation, by which a great many "changes may be rung," amounting in one lock, it is said, to two million. The permutation principle, however, is not contained in the key.

Messrs. Gray and Son's may also be called a "permutating" lock, the construction admitting of infinite variations; in the specimen exhibited, to which the prize medal of the Royal Scottish Society of Arts was awarded in 1850, there are two bolts, each operated upon by a different set of "players;" the lock has four players for one bolt, and three for the other; and the distinct positions in which these players may be placed, by fine workmanship, is said to be 30 for each player; and the committee of the Society report that, allowing 30 distinct positions for each player, the number of different locks which might be constructed would be upwards of 22,540,000,000,000,000,000,000,000, thus affording security against false keys; these locks have at present only been made for bank safes, in which case the size is generally about 20 inches square. A 12-lever lock of this description would cost £7 10s.

Mr. Parke's beautiful padlock is a modification of the Bramah Slider Lock.

The Lever Bolt Safety Lock, exhibited by Windle and Blyth, is very peculiar; in addition to racked tumblers, it has a set of "lever bolt guards." The key is in two connected pieces; the "*nose end*" is provided with a "bit" on one side, and the handle part with a "bit" on the other side; the key, being placed in the key-hole, on being partly turned, assumes a different shape, and in this condition, by one bit, withdraws the "bolt guards," whilst the other shoots the bolt itself

..A peculiar principle is also observable in the key of the lock manufactured by Bryden and Sons, which extends outwards after it is inserted in the key-hole, operating upon distant "players." An essential part of the key is also made removeable at pleasure, without which the key is useless.

The elaborate key-holes observed in some of the old locks are abandoned in those of modern date; and thus, the form of the key being more simple, its imitation is perhaps facilitated, and at the same time the lock is more readily examined by a reflector. A beautifully worked specimen of a key and key-hole was exhibited by Mr. Raab; and this may possibly suggest the idea of greater precaution against the imitation of keys—their *possession* for a time by a good workman being all that is required for a fac-simile to be produced. Such a key as that above referred to would, however, be exceedingly difficult to copy.

Where it is not required that doors should be locked from both sides, there is little advantage in having a key-hole *through*; this only causing the lock to get sooner dirty, by the constant current of air carrying dust into it. The tampering with a key with burglars' nippers, on the outside of a door, is pre-vented by the "safety key," exhibited by Mr. Hanley, the end or point of which is made to turn upon a pin, so that the "nippers" are useless.

Messrs. W. and J. Lea have invented some very useful little key-rests, con-sisting of a brass bracket, about an inch square, to which are attached two pliant pieces of steel, curved to receive the barrel of the key; these may be screwed up in a cabinet or closet, one under the other, and keys fixed and re-moved in a moment, thus admitting of more compact arrangement than the ordinary hooks : prices, according to size, 4s. 6d to 9s. per dozen.

The splendid lock exhibited by Mr. Downes must not be overlooked; its mechanism is peculiarly beautiful, and for strength and security is very valuable ; it is to be placed upon the centre of the door, shooting three bolts from each side, these twelve bolts are secured by "rising bolts," worked by spiral springs, the "rising bolts" also secure four "fly bolts," which are confined by two elliptic springs; the key is in the form of a cross, and, by means of a number of "secret wards," forces the "fly bolts" from their fastenings, at the same time working an eccentric lever wheel, by which the twelve bolts are thrown. The key cannot be withdrawn until the lock is fastened. A lock on this prin-ciple might be also made for bank outside doors, shooting two long bolts, one up and one down. For a door three or four feet wide, and seven high, the price would be £18 or £20.

As a specimen of foreign manufacture, the lock exhibited by M. Bergstrom is worthy of attention, and although of rough exterior, on being opened, the interior work is found to be extremely good. It has eight tumblers, and is without detector : price £3.

Letter Locks.

The principle of these locks is very fully carried out on the Continent; most of their iron safes being fitted with them. In some of these, several alphabets are used, and by this means an almost endless "permutation" is obtained. These locks, when of a common description, are anything but safe. From reposing confidence in the *secret* of the lock, less care is naturally taken of the key, which being improperly obtained, the lock may not unfrequently be opened by attaching a weight to one side of the key, and severally turning round the rings until their correct position is discovered. The continental locks, how-ever, are of most excellent construction, but it is quite possible that if, in haste, a person, after having locked the door, should forget to disconnect the letters of the secret word the "sesame" may be discovered.

This subject must not be concluded without an expression of astonishment and admiration at the exceedingly curious principle adopted in the locks of

Mr. Hobbs' invention, manufactured by Day and Newell, and those of Mr. G. Shmedliér, of Vienna; here the "permutations" are not caused by any variations in the construction of the locks, but are produced by the action of the key. It is pretty well known, that the part of the key corresponding with that which, in other keys, is usually termed the bit, is in these keys formed of a number of flat rings of iron of various sizes,—the relative position of these rings can easily be altered, by removing the end of the key. In the process of locking, the key *forms the lock according to its own shape*, and then no other key will open it. The process by which this is accomplished, is not easily understood, much less easily described; a set of steel bars or plates is operated upon by the rings, which, in the passage of the key, are thrown into certain corresponding positions. The key may be varied almost infinitely, and a fac-simile lock will be produced.

Safes.

For the purpose of awakening attention to the importance of locks and *safes*, it may be suitable here to quote a passage from the *Bankers' Magazine* for April, 1845.

"In a country where a large class subsist by robbery, and where the means of effecting it securely is the constant study of skilful and ingenious thieves, the only means of baffling them, and of protecting the ordinary depositories of valuables from their felonious attacks, are to call in the aid of the greatest mechanical skill with respect to locks and fastenings, and to exercise unceasing care and vigilance. The bank robberies during late years show, that they have been planned with extraordinary sagacity, and have been effected with a degree of skill, which proves that they are not undertaken by ordinary thieves. The large amount of money which the housebreakers are confident of obtaining, in the case of a successful burglary at a bank, induces them to act with a degree of skill and caution proportionate to the expected booty, and it is for this reason, that an unsuccessful attempt to rob a bank is seldom heard of; when "a set" is made at a bank, every information is, in the first place, sought for by the burglars, of the means of security adopted, and it has been ascertained, that many weeks and even months, have been occupied in this manner. Attempts are made to tamper with the servants, and an acquaintance is formed, if possible, with some of the female domestics. If, upon inquiry, it is found that the means of security are so numerous and inviolable as to give no chance of success, the matter is quietly dropped; but if any opportunity presents itself, no time is deemed too long to wait for the proper moment when the bank may be entered, the misnamed *safe*, or *strong room*, be opened, and a clean sweep made of all the convertible securities, and money it may contain."

The discoveries in chemistry have done much to promote security from fire; and the sides of safes, when constructed with highly non-conducting chemical preparations, are capable of withstanding a great amount of heat for considerable periods. Many persons have been surprized at the amazing thickness of the safe exhibited by Mr. Leadbeater. This thickness of nearly 12 inches is not generally necessary, and is intended for cases where, for convenience, it may be placed in the centre of a warehouse containing large quantities of inflammable matter; the sides are filled with non-conductors. A safe, similar to that exhibited, affording 6 cubic feet of clear space inside, viz.,—36 inches high, 24 inches wide, and 12 inches deep, would be £100. Mr. Leadbeater would, however, furnish a safe 3 inches thick at all parts for £75. He has supplied several banks with safes of this description, including London and County Bank, Royal British Bank, Messrs. Walter, Haverfordwest, Mr. Adams, Ware, &c.

E

No one should purchase a fire-proof safe without first reading the interesting pamphlet of fire-proof statistical detail, published by Messrs. Milner and Son, safe manufacturers, with full particulars of their patent principle, which is founded upon the use of chemical non-conductors, together with the use of water, which, in case of fire, by its gradual conversion into vapour, has a powerful influence in resisting heat. The cost of an outside size banker's safe, 6 feet square, and 8 feet deep, would be £200.

The patent fire-proof strong room of wrought-iron is particularly worthy of attention. This is exhibited by Mr. Marr, whose peculiar success in the construction of his locks has been already alluded to. This room, or safe, is 6 feet cube, and the sides, 4 inches thick, are made upon chemical principles of resistance.

Mr. Marr's workmanship is distinguished by stability and care in its execution, and his safes are used by some of the first bankers in the country. One of these was taken red-hot from the ruins of the Royal Exchange, in 1838, when the documents contained were found in a state of perfect preservation.

Foreign Safes.

The fire-proof safe of Mr. Kolesch from Stettin, is singular in appearance, no key-holes being visible, and the locks are so arranged that they cannot be opened, with the right key, by any one unacquainted with the secret. These *secrets* appear generally objectionable on the ground, that great reliance may be placed upon them, whilst all the while, perhaps, their existence is known and understood, although the party interested may not be aware of it. The price of this safe is £82 10s.

A beautiful iron safe is exhibited by Mr. Arnheim, from Berlin; this is especially adapted for a *money* safe, on account of its great strength. It is fitted up with iron drawers.

The spendidly burnished safe by Sommermeyer and Co. is very attractive and finely executed, as are also those of M. Verstaen and M. Paublan, from Paris. In these the fireproofing consists in the side being constructed of double iron with wood between; this, as a slow conductor, is very well, although not equal to the chemical combinations in English safes. The price of the two safes exhibited by M. Verstaen, are £80 and £140.

Wilder's Patent Salamander Safes seem to possess an equal notoriety in America with Milner's safes here. A very long list of testimonials is presented in their favour, and, on many occasions, they have been subjected to severe fiery tests. The price of the small specimen exhibited is £40.

Locks, and especially *safes*, being considered as amongst the most important articles in the Exhibition connected with banking, will, it is thought, justify their having been thus somewhat lengthily discussed; it must, however, be observed, that the "safes" of the Great Exhibition, *as a whole*, are distinguished rather by ornament and beautiful workmanship than by stability and practical utility for banking purposes; they are too small and they are too handsome; and, as a consequence, they are (proportioned to the accommodation afforded) far too costly.

Chemical compounds, which are prepared with great labour and expense, are extremely valuable for safes when these safes are liable to be exposed to fire; but when the property to be secured is so valuable as in banks, it may be desirable to incur some slight inconvenience to place it in a *position* which is in itself fireproof. It is therefore suggested, that bankers' safes, especially money safes, should always be *under ground*. If this cannot be arranged, then they should have fire-proof floors above and under them, and be so placed, that a passage or space should run round or about them, and that no *wooden* fittings or furniture occupy the intermediate space. The safe wall should, of course, be constructed of stone and iron.... The next wall to it might be built of hollow bricks, or these hollow bricks filled with sand. A safe by this means must be entirely fire-proof, and if arranged at the time of building, might be but a slight expense. In case such a room was proposed to be built, as an addition to a bank, the floors being already made, a security from fire might be obtained by adopting Mr. Machay's invention. This consists of "*fusible heads*" attached to the ends of pipes leading from a cistern of water, or a constantly charged main: the pipe is made to terminate in several of these sealed "*fusible heads*" which may be placed in any desired position. In case of fire, if the heat is raised to a certain height, these heads are fused, and the water immediately gains vent and deluges the surrounding parts. If for the mere security of *books*, the stone walls might be dispensed with, and brick used, as it seems hardly necessary that a book-safe should be made specially thief-proof. The strong room just described will be principally suitable for those books of a *valuable character* which have been filled. These *must be kept dry*, which is not easy in the case of papers confined for years under ground, except by constant fires; and although Mr. Clifford's valuable invention for restoring books and papers may do *much*, it is a better plan to *prevent* the necessity for its use. Although the "strong" room mentioned above may be deemed a sufficient security for even valuable books *when filled*, and on account of its superior dryness be preferred, it seems that books *in use* should be kept beyond the possibility of fire, as much as bills and bank notes. These, therefore, should go below, and for this purpose, it is suggested that a cellar room or well, lined and arched with Ridgway's *hollow* bricks, should be constructed, and for convenience, as nearly in the centre of the bank as possible, having a square opening at the top, some distance underneath the floor. That a cubical frame, or cage, of flat bar iron should be formed to correspond in size with the opening, a powerful tackle and windlass being fixed above it. This "cage" might then, by means of "sliding bars," be raised and lowered through the opening into the room below. One side of this cage should be open, into which, when drawn up level with the floor, a barrow might be wheeled, containing the books in daily constant use. The three other sides might constitute a series of book-shelves, externally accessible for those books most frequently referred to by clerks in its immediate vicinity. If the hollow bricks were not found sufficient to keep the room from damp, a stream of hot air from a flue might be occasionally driven through it.

We have thus grouped together such of the productions of human skill, exhibited in the Crystal Palace, as appear peculiarly subservient to the purposes for which this essay was originated.

While diligently engaged in this selection, and testing, in some degree the relative merits of various designs, we have often been astonished at the wonderful works of man, and irresistibly led to the conviction, that "his God doth instruct him to discretion," and while devoutly echoing the very appropriate motto of the general catalogue,—"The earth is the Lord's, and the fulness thereof," to acknowledge "This also cometh from the Lord of Hosts, who is wonderful in council and excellent in working."

A PRACTICAL TREATISE ON BANKING. Fifth edition. By J. W. Gilbart, F.R.S. 2 vols., 8vo, price 24s. Longman and Co.

THE HISTORY OF THE BANK OF ENGLAND, ITS TIMES AND TRADITIONS. By John Francis. 2 vols., 8vo, price 21s. Longman and Co.

A HISTORY OF BANKS AND BANKING. By W. J. Lawson. 1 vol., 8vo, price 16s. Bentley.

THE PHILOSOPHY OF JOINT Stock BANKING. By G. M. Bell. Price 3s. Longman and Co.

THE CURRENCY QUESTION (price 2s. 6d.), and

THE COUNTRY BANKS AND THE CURRENCY. Price 4s. An examination of the evidence on Banks of Issue, given before a Select Committee of the House of Commons in 1840 and 1841. By G. M. Bell. Longman and Co.

These works include a summary of the evidence of Messrs. Cobden, J. H. Palmer, S. J. Loyd, T. Tooke, V. Stuckey, J. W. Gilbart, P. M. James, A. Blair, R. Murray, &c.

LETTERS ON THE INTERNAL MANAGEMENT OF A COUNTRY BANK. By Thomas Bullion. Price 5s. Groombridge.

THE ANATOMY OF COUNTRY BANKS. By James Strachan. Price 2s. Groombridge.

THE BANKER'S CLERK, comprising the Practice and Principles of Banking. Price 1s. 6d. Charles Knight.

A PRACTICAL TREATISE ON THE LAW OF BANKERS' CHEQUES. By George John Shaw. Price 6s. Groombridge.

LOGIC FOR THE MILLION. By a Fellow of the Royal Society. Second edition, price 6s. Longman and Co.

"To all who desire a clear, common-sense, and eminently practical system of Logic, and do not object to the volume that contains it comprising, also, a most entertaining series of extracts from some of the ablest pieces of modern argumentation, we heartily commend this 'Logic for the Million.' We know not where a young man desirous of self-cultivation could more certainly or more pleasurably find it than in this volume. He will only have himself to blame if he does not rise from its study with clearer thought, invigorated powers, and a mind enriched by some of the best good sense of our best writers."—*Weekly News.*

"No young man, desirous of improving himself in the art of reasoning, could more profitably employ his leisure hours than in studying this volume; indeed, we heartily commend it for general perusal."—*Oxford Journal.*

"We hope that 'Logic for the Million' will be read by the million; it will advance their knowledge and improve their taste, their style of writing, and their skill in reasoning."—*Economist.*

"By the production of the first really popular work on a subject of no mean importance, our F.R.S. has added to his own laurels, and conferred a signal service upon the whole community."—*Morning Post.*

INDUSTRIAL EXHIBITION.

———◆———

THE following Extracts from Speeches and Writings referring to the Industrial Exhibition of 1851, are introduced as "ILLUSTRATIONS OF REASONING," in "Logic for the Million—a Familiar Exposition of the Art of Reasoning, by J. W. Gilbart, F.R.S."

I. Extract from an Address delivered by the Earl of Carlisle, at the Meeting at Westminster :—

"The object of the undertaking was too generally known to call for explanation, and too generally approved of throughout the country to stand in need of defence. He was not a member of the Royal Commission to which the office had been entrusted of superintending the management and execution of the project, and he was therefore not called upon to enlighten them with reference to the specific details. He could only make the most general remarks, and submit the most general grounds, which, as he conceived, this undertaking put forward for their support. The first ground was one of a very general character, because it rested on the constitution of our common nature. He liked the occasional recurrence of celebrations, pomps, festivities, commemorations—call them by whatever name you please. They seemed to him in accordance with the feelings and instincts implanted in the human breast. He did not call to mind any country or era which had dispensed with them. In the primitive East, both in ancient and in our own times, they were, for the most part, associated with ceremonies of worship, with uncouth, and too often horrid rites. In more enlightened and civilized Greece, they consisted of public games, in displays of physical strength, ennobled, it was true, by the susceptibility of her people to all the forms of taste and beauty, and recorded in immortal song. Among the sterner Romans they generally accompanied the long train of martial triumphs up the steep ascent of the Capitol, or the more degrading spectacles which summoned the men, and, alas! the women of Rome, to witness the dying throes of the gladiator, or the bloody struggles of captives and wild beasts. In our own age and country they naturally wore a softer aspect; but still, in his judgment, had been too much confined to the easier and wealthier classes, or connected with the pursuits of frivolity and dissipation. It seemed, therefore, but natural and becoming at the period of the world at which we were arrived, that industry, that skill, that enterprise should in their turn have their own ovation, their own triumph, their own high holiday, where the workmen and workwomen of the world might enjoy a day's pause from their engrossing toils for the purpose of seeing what their fellow-workmen and workwomen were doing and could do all the world over—that they might see, not barbaric rites, but useful inventions; not exhibitions of physical prowess in the prize-ring, or the foot-race, but results which interested the mind, and elevated the soul; not suppliant provinces and chained captives, but the pursuits of peace and of civilization— not crowded saloons, and heated theatres, but an arena where all ranks might mingle, where all might learn, and all might profit by what they saw. This was one of the grounds which the undertaking put forward for support. It might be reckoned one, perhaps, rather of a sentimental and fanciful character; he would proceed to one somewhat more practical. The Exhibition, carried into effect on the scale proposed, would give people in all pursuits and professions, in all classes and callings, the opportunity of examining and ascertaining methods by which the work which formed the daily business of their lives might be expedited, facilitated, assisted, and improved. There might be the textures which best suited the climate in which we lived. There might be the tools and implements calculated to lighten or shorten labour. There might be the discoveries in mechanics which should mould to the purpose of man the unvarying attributes of matter, space, and time ; discoveries of which we, perhaps, had long been in the unconscious need, but which had long been beneficially adopted in other countries. He could not but look forward with pleasure and hope to the evidences which our countrymen would afford of the distinguished progress they had made in the pursuits of civilized life, and especially in those which interested most permanently the well-being and comfort of the millions. And though he could not forget, on this occasion, as he could not forget on any occasion, his long connexion with a district which had attained a marked eminence in all these departments of production—the West Riding of Yorkshire, the region of our cutlery, of our hardware, of our great woollen and worsted manufactures—yet he must say, that he not only anticipated advantage to the country from our victories on this occasion, but also from our defeats. It was only when we could compare and put side by side what we could not do and what we were not, with what we could do and with what we were, that we could attain the true measure of our superiority, and of our deficiencies. Nations, equally with individuals, should say, after Brutus,—

'I should be glad to learn of better men.'"
—*Logic for the Million*, p. 290.

II. Extract from a Prize Essay written by Mr. Briggs, a working millwright :—

"The working man, let him be engaged in whatever kind of labour or handicraft, by attending the Exhibition, will find information, valuable and essentially useful to him. The mechanic, and the worker in metals, will be able there to trace the mineral as it appears when first dug from mother earth, through all its varied processes, till finished in the powerful and almost intellectual steam engine. The workers of textile fabrics will there be able to inspect the silk as produced from the

silk-worm, and the cotton from the plant, tracing their progress until, by the aid of man and machinery, they appear in
the finished fabric, exhibiting the greatest beauty of design
and skilful workmanship. The workman, at this Exhibition,
will be able to compare the textile and other productions of
France, the marble productions of Florence, and specimens of
art and science from all parts of the Continent, with those
exhibited by our country, and perceive that the respect in
which our continental neighbours excel us is in fertility and
beauty of design. He will see that no nation executes so well
as the English ; that we stand unrivalled in the superiority of
our workmanship ; and that the only thing we are short of is
design. And if a visit to the Exhibition conduce to the cultivation of a superior style of·design among our artisans, it
will repay a hundred-fold any expense they may incur by
attending it, and confer lasting benefit on their country, for
the acknowledged character of the British, as the most skilful
of workers, added to that of superior designers, would always
demand for us a trade which would proudly and successfully
outstrip every other nation in the world in the race of competition. The history of the marble trade in my native county
of Derbyshire, is a striking illustration of the truth of this.
Twenty or thirty years ago the art of inlaying the marble was
not known there. And why is it known now ? Because the
museums of Matlock and Buxton imported from Florence,
Rome, and other parts of Italy, specimens of the art, which
were seen and imitated by the workers of Derbyshire ; until
at length they are able to equal, and, in some respects, surpass
the productions of Florence. And if the small exhibition of
Matlock has been the means of producing an entirely new
and beautiful art amongst the inhabitants of the Peak of
Derbyshire, what may we confidently expect to result from
this great and mighty Exhibition, which, in comparison to
that at Matlock, is like the broad Atlantic to a gentleman's
fish-pond. The great want of our country is design ; we have
no other want that prevents us from successfully competing
with foreigners. Then let our artisans attend this Exhibition,
and ascertain where and in what they excel them ; and it will
propel the current of improvement at home, and thus minister
to their advantage, and to the comfort and happiness of the
whole people."—*Logic for the Million*, p. 90.

5

III. Extract from the speech of Lord Overstone, delivered in the Egyptian Hall, Mansion House :—

"It was no longer necessary for them to discuss its character or expediency; on these points the public were united. The scheme now went forth to the world stamped by the approbation and recommendation of *our gracious Sovereign*, as expressed in the act constituting the Royal Commission, and as further proved by the munificent donation with which she had headed the subscription. It was further stamped with the continued support of *her illustrious Consort*, whose original suggestion it was, and who had already been alluded to in terms not more complimentary than he deserved; for those who wished to know his true character should see him seated at the council board, and observe his capacity for conducting complicated public business with sagacity, perseverance, and energy. Beyond this the project went forth sanctioned by the authority of *the Ministers of this country*, as evidenced by their advising the Crown to issue the Royal Commission, and by their presence on this occasion, through which so much weight was given to the proceedings of the day. It was still further sanctioned by the *recorded resolutions of the people of England* assembled in almost every great district of the empire, and lastly, by the *two great meetings held in the Egyptian Hall*."—*Logic for the Million*, p. 61.

IV. Extract from an article in the *Times,* upon the effect of the Industrial Exhibition with reference to the London shopkeepers:—

"In what way can the Exhibition have injured trade? Let us place the matter steadily before us for consideration, and ask ourselves by what conceivable means the vast addition that has been made to the floating population of London can possibly take money out of the pockets of our tradesmen? Here are the railway termini and steamboat quays daily pouring upon the town their thousands of visitors to the Crystal Palace. How can the presence of these almost innumerable multitudes be said to injure trade? Granted that considerable bodies of the labouring population come and go in the same day, even these have given a considerable stimulus to the receipts of the railways, and of the ordinary conveyances which ply about our streets. Thousands no doubt come and go, but thousands remain for twenty-four, for forty-eight hours, for a week. About 50,000 persons daily visit the Crystal Palace. We presume that these sojourners are affected by the ordinary wants of humanity. They must eat something, they must drink something, they must sleep somewhere. It is certain that the great bulk of the number will soon have returned to their homes, yet they will be replaced by others, who will be replaced by others in their turn. Now, is it natural to suppose that the ladies of the different parties, old or young, will be indisposed to carry back from 'town' such articles of finery as may be best calculated for the enslavement of the country-side ? We will not confine the observation to ladies alone, for we have yet to learn that men who come up from the provinces to make holiday, with a reasonable amount of ready money in their pockets, do not usually succumb to temptation before the brilliant shop-fronts of the metropolis. A watch must be purchased for one fair cousin, a dress for another, and so on. Each lays in a stock of presents for absent friends, according to his means and degree—the selfish fellows buy for themselves. We are surely not exaggerating matters, for we find, on turning to the traffic returns of the railway for the past week, as compared with the corresponding week of last year, that the London and North Western has taken 66,561*l.* as against 51,492*l.* Now, all due abatement made for other causes which might have stimulated traffic, is there not here enough to show the nature of the addition that must have been made to the floating population of the metropolis during the past week ? The past week is merely a repetition of other weeks that have preceded it, and a sample of others that are to follow.

" We have not dwelt upon the influx of foreigners that has actually taken place, although a considerable increase in the usual number of our continental visitors must, of necessity, exercise a very important influence upon trade. Granting that the expenditure of each of them individually may be but slight, yet taken collectively, that expenditure must represent a vast amount of consumption, and consequently of profit to the London tradesman. Common sense, then, would tell us *à priori* that a great concourse of people within the limits of a particular town must exercise a very beneficial influence upon trade. When we come to try the case by other tests, how does it stand? Were the streéts of London ever more crowded than during the present Exhibition year? Are they not swarming with carriages, and throngs of well-dressed people? Is there not more than the usual difficulty in getting served in any shop which has attained a celebrity in any particular department of trade? Has the 'season,' as it is called, been a bad one for those tradesmen who minister to the luxurious tastes of the fashionable world? If you passed at night through any of the quarters of the town in which Fashion holds her seat, you might see mansion after mansion lighted up, and find to your sore discomfort that the usual strings of carriages 'stopped the way.' A bare reference to the columns of the public journals in which such matters are usually recorded would be sufficient to show that never, as far as entertainments go, has a gayer season been known. Of all the ill-timed and illogical cries that ever were raised, this one of injury to the trade of London in consequence of the Great Exhibition is the most fatuous and the most untrue. We should be curious to have in a succinct form a detailed account of the miseries of some London shopkeeper who could establish any connexion between his losses and the Crystal Palace in Hyde Park. Still more should we feel obliged to any one who could favour us with any approximation to the *per contra.*"—*Logic for the Million,* p. 292.

V. Extracts from a Prize Essay, written by the Rev. Mr. Whish :—

" Even with reference to the parts of any single nation, it is *the lack of facility of intercourse* which is the acknowledged *cause of all that is defective* in the rural population. It perpetuates peculiarities of idiom and of pronunciation, local prejudices, inactivity of mind, roughness of manner, and subjection to the power of superstition. Everything, therefore, which quickens circulation or facilitates intercourse between either the different members of the same nation, or between members of different nations, is calculated to promote the general welfare."—*Logic for the Million,* p. 81.

"Let us consider some of those points in which other nations offer us a high example. We may mention, for instance, that there is among the continental nations a *general amenity of manners,* a freedom of intercourse between the various classes of society, which certainly gives them the appearance of great amiability, besides that it is the source of other advantages. Again, we find in the nations that belong to the Roman Catholic Church, a straightforward, unaffected *boldness in the profession of their religion,* which is worthy of a purer creed. There is also in Roman Catholic countries a *regard to the outward forms of religion* which, though not in itself all that is required of the Christian, nor even the most important part of his duty, is yet the natural manner in which a real spirit of religion should exhibit itself. It is, however, in the eastern world that religious feeling is exhibited in the most natural manner. We may perhaps have among the approaching throng of interested spectators some of the followers of Mahomet, whose well-known custom it is frequently to *ejaculate their brief confession of faith,* and who would never think of writing a book without prefacing it by an inscription of praise to God. We shall doubtless have a close criticism instituted upon *our mode of education,* and inquiry as to the degree in which it meets the wants of our population. The inhabitants of those countries where attendance upon schools is compulsory upon children of a suitable age, or of those in which it is universally adopted from a real estimation of the benefits to be derived therefrom, may perhaps be surprised at the defects and imperfections which are allowed to exist in our system. The ample provision which has been made in many of the American States, for this purpose, at a very early period too after their first establishment, deserves to be noticed as affording an example most worthy of imitation. We might go on in the same strain and speak of that *high sense of filial duty,* which the disciples of Confucius would expect to find in us, and which is among themselves an effectual principle of government, or of that *tenderness towards dumb animals,* and that *strong feeling of brotherhood* pervading the different sects, and superseding all necessity for poor-houses, which are so generally manifested by the worshippers of Brahma, and which may therefore be considered by them as the best evidences of moral excellence."—*Logic for the Million,* p. 155.

" The many friendships that will be established during the existence of the Exhibition between the members of different nations, will be so many powerful motives for resisting war, so many guarantees for quiet and reasonable legislation ; the breaking down of unfounded prejudices, a more accurate and enlarged knowledge of the real character of our neighbours, the right appreciation of their talents and other excellences, the perception of those points in which we ourselves are inferior to them,—all these things have the same tendency, and they may rationally be expected to follow from that more close collision with foreigners which will be caused by the Great Exhibition of Industry. It is not enough, therefore, to say that it will, under this aspect, promote the welfare of mankind ; we may boldly say, it will promote their *moral* and *religious* welfare."—*Logic for the Million,* p. 357.

VI. Statistics of the Industrial Exhibition of 1851 :

" The 'CRYSTAL, OR GLASS PALACE,' prepared for the 'World's Fair,' or great industrial exhibition of 1851, is 1,848 feet long, by 456 in width. The height of the three roofs is 64, 44, and 24 feet ; and that of the transept, 108 feet. The ground floor occupies 752,832 superficial or square feet ; and the galleries, 102,528 feet, making, in all, an exhibiting surface of some 21 acres, with a length of tables of about eight miles. There are 3,500 cast and wrought-iron columns, varying from $14\frac{1}{2}$ to 20 feet in length ; 2,224 cast-iron girders, and 1,128 supporters for the galleries. The glass necessary to cover this immense building, is 900,000 square feet ; the length of sash-bars is 205 miles ; and there are 34 miles of gutters to carry off rain water to the hollow columns, through which it passes into drains or sewers underground."

B

The following Return has been compiled from the official lists published daily, and shows the estimated weekly number of visitors, and the money taken at the doors, from the week ending May 3d, to Saturday the 30th of August.

Week ending	Number.	Entrance Fee.	Amount received at the doors.			Estimated number of persons entering with season tickets.	Total No. who enter weekly, including the staff, exhibitors' attendants, and the press.
			£	s.	d.		
May 3	1,042	£1	1,042	0	0	49,000	56,042
10	41,194	5s.	10,298	10	0	77,056	118,250
17	53,386	5s	13,346	10	0	30,121	145,507
24	89,458	5s.	22,189	0	0	91,440	192,869
31	160,857	1s.	11.123	5	0	61,257	222,114
June 7	218,799	—	13,694	2	0	27,129	245,928
14	206,233	—	12,943	12	0	32,352	238,585
21	267,800	—	16,421	3	0	55.215	303,015
28	262,464	—	16,177	8	0	30,245	292,709
July 5	225,503	—	14,073	0	0	21,436	246,739
12	265,319	—	16,427	5	0	23,108	288,427
19	283,400	—	17,516	0	0	22,453	305,853
26	255,768	—	15,761	4	0	18,371	247,139
Aug. 2	270,900	—	16,315	17	6	15,617	288,519
9	266,770	—	15,440	14	6	20,001	286,771
16	236,096	—	14,050	18	6	15,961	252,057
23	226,502	—	13,360	12	6	10,037	236,539
30	202,808	—	11,860	7	6	8,638	211,446
Total number of Visitors from May 1 to Aug. 30							4,205,509

—Logic for the Million, p. 348.

R. CLAY, SON, AND TAYLOR, PRINTERS, BREAD STREET HILL.

THE LONDON AND WESTMINSTER BANK.

THE following Table shows the amount of Paid-up Capital, Annual Profits, Dividends, Surplus Fund and Deposits, of the London and Westminster Bank, on the 31st December in each year, from the opening of the Bank to the year 1859 inclusive :—

Date.	Paid-up Capital.	Profits of the Year.	Dividends and Bonuses.	Surplus Fund.	Deposits.
	£ s. d.	£ s. d.	£ s. d.	£ s. d.	£
1834	182,255 0 0	3,550 6 6	2,334 18 1	1,205 8 5	180,380
1835	267,270 0 0	11,520 10 0	10,818 12 0	1,907 6 5	266,884
1836	597,255 0 0	32,483 14 1	29,864 0 0	4,527 0 6	643,332
1837	597,280 0 0	32,404 10 8	29,864 0 0	7,067 11 2	793,148
1838	597,280 0 0	43,635 12 11	29,864 0 0	20,839 4 1	1,387,855
1839	597,280 0 0	48,098 3 0	35,836 16 0	33,100 11 1	1,266,845
1840	597,280 0 0	48,951 8 10	35,836 16 0	46,215 3 11	1,361,545
1841	786,300 0 0	51,300 0 9	41,507 8 0	56,007 16 8	1,499,328
1842	800,000 0 0	55,118 14 2	48,000 0 0	63,126 10 10	2,087,757
1843	800,000 0 0	51,696 5 7	48,000 0 0	66,822 16 5	2,219,624
1844	800,000 0 0	51,081 18 11	48,000 0 0	69,904 15 4	2,676,741
1845	800,000 0 0	66,344 1 0	48,000 0 0	88,248 16 4	3,590,014
1846	800,000 0 0	72,175 15 9	{48,000 0 0 16,000 Bonus.}	98,424 12 1	3,280,864
1847	988,882 0 0	58,223 4 10	54,000 0 0	100,647 16 11	2,733,753
1848	998,768 0 0	62,076 0 0	60,000 0 0	102,723 16 11	3,089,659
1849	1,000,000 0 0	65,120 17 7	60,000 0 0	107,844 14 6	3,680,623
1850	1,000,000 0 0	67,262 17 7	{60,000 0 0 15,000 Bonus.}	100,107 12 1	3,969,648
1851	1,000,000 0 0	84,044 9 0	{60,000 0 0 20,000 Bonus.}	104,152 1 1	4,677,298
1852	1,000,000 0 0	85,012 6 9	{60,000 0 0 20,000 Bonus.}	109,164 7 10	5,581,706
1853	1,000,000 0 0	116,142 13 8	{60,000 0 0 40,000 Bonus.}	125,307 1 6	6,259,540
1854	1,000,000 0 0	149,219 9 5	{60,000 0 0 80,000 Bonus.}	134,526 10 11	7,177,244
1855	1,000,000 0 0	159,583 8 5	{60,000 0 0 90,000 Bonus.}	144,109 19 4	8,744,095
1856	1,000,000 0 0	176,868 7 3	{60,000 0 0 110,000 Bonus.}	150,000 0 0	11,438,461
1857	1,000,000 0 0	180,196 15 7	{60,000 0 0 100,000 Bonus.}	158,596 10 9	13,889,021
1858	1,000,000 0 0	180,232 15 8	{60,000 0 0 110,000 Bonus.}	181,407 17 10	11,465,815
1859	1,000,000 0 0	208,065 12 10	{60,000 0 0 120,000 Bonus.}	200,000 0 0	11,115,697
		2,160,410 0 9	1,950,936 10 1	200,000 0 0	

Thus it appears that the total profits of the Bank, from its opening on the 10th March, 1834, to December 31, 1859, amounted to 2,160,410l. 0s. 9d. ; the Distribution to Shareholders in Dividends and Bonuses amounted to 1,950,936l. 10s. 1d. ; the Surplus Fund amounted to 200,000l. ; and there remained to the credit of Profit and Loss Account, 9,473l. 10s. 8d.

CPSIA information can be obtained
at www.ICGtesting.com
Printed in the USA
BVHW07s1104210818
525176BV00011B/299/P

9 781331 915331